Grassroots School Reform

Previous Publications by the Author:

A Fieldbook for Community College Online Instructors (2007).

Leadership as Service: A New Model for Higher Education in a New Century (2007).

Grassroots School Reform

A Community Guide to Developing Globally Competitive Students

Kent A. Farnsworth

GRASSROOTS SCHOOL REFORM
Copyright © Kent A. Farnsworth, 2010.

Cover Photo by Sandra Cunningham, 2010. Used under license from Shutterstock.com.

First published in 2010 by
PALGRAVE MACMILLAN®
in the United States—a division of St. Martin's Press LLC,
175 Fifth Avenue, New York, NY 10010.

Where this book is distributed in the UK, Europe and the rest of the world, this is by Palgrave Macmillan, a division of Macmillan Publishers Limited, registered in England, company number 785998, of Houndmills, Basingstoke, Hampshire RG21 6XS.

Palgrave Macmillan is the global academic imprint of the above companies and has companies and representatives throughout the world.

Palgrave® and Macmillan® are registered trademarks in the United States, the United Kingdom, Europe and other countries.

ISBN: 978–0–230–10832–5 (hardcover)
ISBN: 978–0–230–10833–2 (paperback)

Library of Congress Cataloging-in-Publication Data

Farnsworth, Kent Allen.
    Grassroots school reform : a community guide to developing globally competitive students / Kent A. Farnsworth.
        p. cm.
        ISBN 978–0–230–10833–2 (alk. paper)—ISBN 978–0–230–10832–5 (alk. paper)
        1. School improvement programs—United States. 2. Charter schools—United States. 3. Schools—Decentralization—United States. 4. Education and globalization—United States. I. Title.
LB2822.82.F37 2010
371.2'070973—dc22   2010014273

A catalogue record of the book is available from the British Library.

Design by Newgen Imaging Systems (P) Ltd., Chennai, India.

First edition: November 2010

10 9 8 7 6 5 4 3 2 1

Printed in the United States of America.

To my wife, Holly,
whose commitment to great teaching and
to her students' success always give me hope

CONTENTS

ACKNOWLEDGMENTS

In several places in this book I mention Mirra Anson who assisted throughout its development as a research and editorial assistant, idea generator, and cheerleader. I cannot adequately express my gratitude for her tireless work and constant good humor. Thank you, Mirra. I am eternally grateful. A number of people also offered personal experiences, observations, and practices that served as inspiration and example: Derek Urhahn and the faculty and staff at Leopold R-III Schools, Scott Shirey and the faculty and staff at KIPP's Delta College Prep, Steve Baugh, E. Allen Bateman, Richard Clement, Brian Croone, Jennifer Cuevas, Rudy Farber, David Farnsworth, Dean Farnsworth, Gib Garrow, Holly Jacobs, Beth Janssen, Barbara Lontz, Charles McClain, Ken Owen, Juliet Scherer, Terry Suarez, and Jim Tatum. Thanks to each of you for your invaluable assistance. My youngest son Paul generously allowed me to use some of our travel experiences as examples, and his wife Jillian assisted with an attractive and appealing cover design. I am very grateful to you both.

My special thanks to Mary Ann Lee whose generous support to the University of Missouri-St. Louis endowed the faculty position that enabled me to do this work, and to Burke Gerstenschlager and Samantha Hasey at Palgrave Macmillan for their helpful advice, advocacy, and editorial assistance. This was truly a collaborative endeavor.

Introduction

Like many parents, mine thought that I would benefit from playing a musical instrument and arranged for me to take both piano and violin lessons when I was young. I didn't particularly care for either and remember complaining about the burdens of both practice and recitals to a trusted leader of a church youth group, confiding that what I wanted more than anything in the world was to be able to play the guitar like some of the pop music greats of that time.

"Do you have a guitar," he asked?

I told him that my parents had given me one for Christmas—a very nice acoustic guitar that I really liked.

"Do you practice much?"

"Yeah, I get it out sometimes and practice some chords. But when the weather's good, I'd rather be out shooting baskets."

"Then I would guess that it's not what you want more than anything else in the world," he said. "I think that what we *really* want the most in the world are the things we spend our time and energy trying to get."

If we were to ask parents what they want more than anything else for their children, most would probably say that they would like them to be happy, healthy, and financially secure, with satisfying personal and perhaps spiritual relationships in their lives. But, for many, these "greatest wants" are a bit like my wish to be an accomplished guitarist—something they commit time and effort to only when a myriad of other distractions don't pull them away. They do own the guitar—the instrument that could provide a number of these greatest wishes—but they choose not to commit the time and energy needed to make the wishes a reality.

In an Op-Ed article in the *New York Times*, columnist and futurist Tom Friedman recounted a conversation with Todd Martin, a prominent business executive-turned international investor, during which Martin explained why the recession of 2008–2009 was about much more than a collapse of Wall Street and the housing market. "Our education failure is the largest contributing factor to the decline of the American worker's global competitiveness, particularly at the middle and bottom ranges," Martin argued, then explained that our loss of competitiveness has compromised our own productivity at the exact time technology has heightened international competition. Our standard of living in the United States has been based on borrowing and consuming beyond our incomes, so when the recession wiped out credit and devalued our major assets, many of us found ourselves unemployed or underemployed and lacking the skills to get back into the new globally competitive marketplace.[1]

Friedman goes on to explain that the "New Untouchables" in the postrecession job market will be those who have not only a solid secondary and college education, but training that assists the student in developing a sense of "entrepreneurship, innovation and creativity." The bottom line, he argues, is that there will be no complete postrecession recovery "without fixing our schools as well as our banks."[2] And therein lies the ability held by all parents to give their children the gift of a promising future—their schools. While parents working collectively within their communities cannot fix Wall Street, they can fix their schools. And if what they want more than anything for their children is economic success, health, and personal well-being, this is where they will focus their attention.

They may dream of Junior becoming a running back in the NFL or of Missy starring on the pro tennis circuit and prefer to dedicate all of their extra time and attention to supporting local high school athletics or the youth tennis and football programs in their communities. But the chances of a high school athlete becoming a player in the NFL are nearly a million to one, while the odds are even for the likelihood of financial success for a student with a strong secondary and college education who has developed a spirit of entrepreneurship, innovation, and creativity. Add fluency in Arabic, Farsi, or Mandarin Chinese, and the future is virtually assured. And while no parent can guarantee that spot in pro athletics to her child, with the right degree of commitment and the right relationship with the community's schools, most parents can assure their children a good education.

But to provide our children with a *great* education—with the globally competitive education Todd Martin was discussing with Tom Friedman—our schools need to be fixed. And, here again, parents who want a bright future for their children more than anything else in the world can make that happen. This book is about how it can be done. It explains that great schools are products of their communities—of the families who live within the districts and select the school leadership. It demonstrates that great schools are not a product of huge financial investment, but of huge commitment—commitment to good and courageous leadership, to strong curriculum, and to high expectations. If we wish more than anything else in the world to create a bright future for our children, the power is in our hands, and we can reform America's schools by rescuing the one in our own community.

CHAPTER ONE

We Need School Reform—and Soon!

If we continue on our current course, and the number of nations outpacing us in the education race continues to grow at its current rate, the American standard of living will steadily fall relative to those nations, rich and poor, that are doing a better job.
—Tough Choices or Tough Times. p. xix

This book was written as much for myself as for anyone else. For almost a decade before beginning it, I had been fighting a gnawing belief that education in America—particularly elementary and secondary education—was in a death spiral that would inevitably lead to disaster. I was spending several weeks each year working with colleges in various parts of the country that were struggling against the effects of a failing system, and were desperate to find ways to make up for those deficiencies. But the evidence was everywhere and was irrefutable. Thousands of young people were graduating from high school seriously underprepared for both work and further education. Some were awarded high school diplomas that were indistinguishable from the class valedictorians, but were based on the student having met the requirements of an Individual Educational Plan (IEP) that was purely behavior-based, and did not require mastery of any body of knowledge. Worse still, a quarter of ninth graders were not finishing school with their classes at all, and of those who did graduate, a high percentage was deficient in reading, math, and English skills. Many who entered college did not survive the first year—in an economy where we were being told that 90 percent of the fastest growing jobs in America required some additional education beyond high school.[1]

In my own home city, St. Louis, Missouri, the mayor hired a not-for-profit group to assess which urban schools were doing the poorest job and might be closed, to be replaced by what he hoped would be higher-performing schools. The evaluation compared school performance on state standardized tests and found that only one neighborhood school in the entire city met state testing standards for the year of the assessment. The evaluating agency had intended to use these pass rates as a measure for identifying "good schools," but with practically none meeting the standard, was forced to create a measure of its own. The new criteria for being classified as "acceptable" required one-quarter of the students to test at the proficient level in English and math. Even so, when the city was divided into areas by zip code, five areas containing a combined population of over 130,000 people had no public schools, including several charter schools, that met the criteria.[2]

At the same time, I was involved in a development project with colleges in Southeast Asia where it was evident that educational systems were becoming better and better—and where there was a visible hunger on the part of students to learn and achieve. During one of those years, a young Vietnamese English teacher lived with my family in the United States while gaining certification in Hospitality Management at a local community college. She was teaching English to students in a hospitality program in Vietnam, and wanted to increase her familiarity with the principles and vocabulary of the discipline being pursued by her students. While at the community college, she set the curve in all of her classes each week and reported to us daily on how she had done on assignments and tests. If she missed a test question, she was visibly disturbed and went over the material again until confident she had mastered it.

After a major midterm exam, she came home one afternoon and reported sheepishly that she had missed two questions. "But I'll do better next time," she promised. I pointed out that she was leading the class on every test and every assignment, and that it didn't matter if she missed a question or two every now and then. My comment seemed to surprise and disappoint her.

"It matters to me," she said.

These visits to elementary, secondary, and college classrooms in countries throughout East and Southeast Asia showed me that how students do in school matters very much to tens of millions of families in that part of the world. Each knows that to have any chance at all of getting into college, children must do well on a grueling battery of exams that signal the end of secondary school. But mastering any particular body of knowledge doesn't seem to matter much at all to those in classrooms I visit in the United States. Here, many students seem either unaware or unconcerned that we have lost our standing as the best-educated population in the world, and are rapidly giving up our position as the dominant economy. They seem either oblivious to the fact or don't care that the major reason we are losing both our manufacturing and our professional employment base to other countries is because these countries have better-educated workers, who will do the work for less money. National data show that no matter what we try at the state or national level, educational performance is not improving, and I keep thinking of Lily Tomlin's line that "No matter how cynical you get, it is impossible to keep up." Education in the United States seems irreversibly doomed to failure.

Then during a visit to a college I was working with in eastern Arkansas, I heard about a small charter school that was showing remarkable results. I made an appointment to visit and discovered that education in America *can* succeed, and that it can succeed with students who have traditionally been viewed as having the greatest challenges. The remarkable thing about this school's success was that there was nothing revolutionary about what it was doing. In fact, it was more a case of being reversionary—of going back to old practices and old expectations. The school looked very much like some that I visit in Asia—with students in uniforms sitting attentively in class, showing respect for their teachers and for each other, and demonstrating the same hunger for learning.

I came away from that visit mildly depressed and discouraged that we can't manage to duplicate this same success in schools in every community across the country. But during my five-hour drive from that small Arkansas community to my home in St. Louis I realized, like being hit between the eyes by the proverbial two-by-four, that we could. The remarkable thing about our system of education is that we are free to make it as strong or as weak as we choose—and we demonstrate this truth in towns and cities across America every day. We like to attribute the differences that exist in our schools to resources, to our assessment that some schools are better funded, better equipped, and have students who come from more advantaged backgrounds. But this small school in eastern Arkansas demonstrates that none of these factors is the critical one. It has modest resources, meets in temporary trailer buildings, and serves one of the most disadvantaged populations in America. The difference is that it *chooses* to be excellent. It *chooses* to have expectations for its students that demand exceptional performance, and it *chooses* to have a class schedule, a school calendar, and a faculty selection and evaluation process that foster achievement.

The epiphany that occurred during that drive to St. Louis was that every school in every community in America could *choose* to be just as successful, but it will have to be done community by community. This book explains why that transformation must happen if we are to remain a strong, secure, and competitive nation, and how it can be done. The unfortunate truth is that we figured this all out once before but didn't choose, as a nation, to do anything with what we learned.

A Nation at Risk

In April of 1983, the Department of Education in Washington DC under the Reagan Administration issued a report that sent ripples across the country and particularly through the education community. Its opening line was "Our Nation is at risk," and in the introduction the report declared that "if an unfriendly foreign power had attempted to impose on America the mediocre educational performance that exists today, we might well have viewed it as an act of war."[3] The remaining pages of the *A Nation at Risk* study explained that major changes in the international economy were reflections of redistribution of skilled workers around the globe, and of new freedom to move work to where these employees live.[4] The report ominously asserted that:

> The people of the United States need to know that individuals in our society who do not possess the level of skills, literacy, and training essential to this new era will be effectively disenfranchised, not simply from the material rewards that accompany competent performance, but also from the chance to participate fully in our national life.[5]

More simply stated, we were being warned that if we did not provide better education for our children, they would not only miss many of the economic advantages enjoyed by their parent's generation but would not be in a position to effectively change their circumstances.

In a series of twenty-nine findings, *A Nation at Risk* outlined the shortcomings of a system of education that it described as having "lost sight of the basic purposes of schools, and of the high expectations and disciplined efforts needed to attain them."[6] It recommended that high school graduation requirements be strengthened and made more rigorous with longer school days, extended school years, and greater emphasis on English, science, mathematics, and foreign languages. Teachers needed better preparation and training, along with better pay, and we needed more teachers in math and science. In essence, education needed a complete overhaul if students who were entering school in 1983 and would be graduating in 2000 were to be prepared to meet the challenges of the new century. Perhaps most importantly, the report asserted that public school boards and community leaders should be held responsible for insuring that these reforms occurred, and higher education institutions should raise admissions requirements to force improvement at the primary and secondary levels.

In the small Iowa community where I was living at the time, *A Nation at Risk* grabbed newspaper and television headlines for days. Virtually everyone knew about its doom and gloom predictions if things did not improve, and about the radical changes the report claimed were needed. In downtown shops and cafes there were voices of concern and consternation, defensiveness and denial. We were all certain that school performance must be pretty bad somewhere else—possibly across the river in Illinois or out East, like in New York—since we knew it wasn't that bad in our town. Surely, wherever the situation was that serious, someone would do something about it, at least we hoped so. We went about our business without doing much at all and, apparently, so did the rest of the nation. After a period of initial hoopla and finger-pointing, our educational system slogged along pretty much as it always had, with one notable exception.

Assessing Stagnation

In state legislatures and in departments of education across America, Item 3 of the *Nation at Risk* Report under Recommendation B, *Standards and Expectations*, set off a ripple of activity. This item suggested that "Standardized tests of achievement (not to be confused with aptitude tests) should be administered at major transition points from one level of schooling to another." While ignoring the recommendations for substantive change, America took this "testing" recommendation to heart and a nation at risk became a nation of testers. States developed systems of assessment that measured student achievement at key grade levels and at the completion of high school. Community colleges introduced entering assessments that determined whether students were ready for college-level work, or needed remediation. Even the federal government stepped into the act, insisting that there be "No Child Left Behind" and mandating that if states and schools were to receive federal aid, testing requirements must be in place to determine if schools were meeting the challenge.

What did all of this testing demonstrate? It proved in graphic and well-documented detail that, over the twenty-five-year period that followed publication of *A Nation at Risk*, practically nothing improved in terms of the academic performance of our students. A new set of national reports published during the first decade of the twenty-first century only served to validate our failures, and to tell us that the risks were real. We are falling dangerously behind the rest of the developed world in educational achievement and economic readiness, and are targets of much of the developing world that sees our employment base as overpaid, increasingly underprepared, and ripe for the picking.

Through the 1970s—and in fact into the 1990s—we could proudly boast that we were Number One on the international charts in the percentage of our population with a high school diploma and college degree. Then all progress seemed to grind to a halt, as other nations aggressively passed us by. The report of the Spellings Commission on the future of higher education published in 2006, *A Test of Leadership*, announced that in a period of only three or four decades the United States slipped to 12th position in higher education attainment and 16th in high school graduation rates.[7] This was not so much a case of our moving backward as one of our standing still, mired in a self-created educational morass, while other nations placed increased emphasis on academic achievement and job readiness.

Economists Eric Hanushek and Ludger Woessmann looked at the relationship between educational achievement and economic development by combining data from thirty-six international tests of academic achievement administered between 1964 and 2003 in fifty countries. Their data provide compelling evidence that economic growth in the latter half of the twentieth century correlated directly with the academic performance of each nation's children.[8] In an Op-Ed piece in the *New York Times*,

Nobel Prize winning economist Paul Klugman commented that "If you had to explain America's economic success with one word, that word would be 'education.' "[9] This observation has a sobering counterpoint in that it suggests that if you had to explain America's decline in economic success with one word, that word would also be "education."

Yet despite this understood relationship between education and economic vitality, right now in the United States one out of four ninth graders drops out of school before the end of the senior year—a statistic that has not changed appreciably over the past quarter century.[10] A report by ETS[11] entitled *America's Perfect Storm* notes that high school graduation rates peaked at 77 percent in 1969, fifteen years before the *Nation at Risk* study, then dropped to 70 percent by 1995, and have remained near that level since. More importantly, the graduation rate of the country's disadvantaged minorities is estimated at 50 percent,[12] and in some of our large urban high schools, fewer than four out of ten freshmen complete high school.[13] In an online review, the school that earned the dubious honor of being America's worst high school had a graduation rate of 7.5 percent the year the study was conducted.[14] And this is happening when virtually all of the population growth in the next several decades will be in the demographic groups that show the lowest rates of educational achievement.[15] Scores on the SAT exam have remained almost constant since the mid-Seventies—averaging 1026 in 1974 and 1026 in 2004. Granted, there has been some shift during this time in categories, with the verbal score falling from 521 to 508, and the math score rising from 505 to 518.[16] But by practically any measure, since the *A Nation at Risk* report was published, very little has improved in American education.

The Storm Arrives

Why is this more critical now than it was in 1983 when *A Nation at Risk* was published? Putting it simply, it is more important today because our economic future as a nation depends on being better-educated, and in dramatically different ways. *Time* Magazine conducted a revealing interview in January of 1990 with the late Peter Drucker who many consider to be the father of modern management practice, in which he spoke about his concerns for the approaching new century. The interviewer, Edward Reingold, began by asking:

"In the remaining years of the 20th century..." at which point Drucker interrupted and said: "We are already deep in the new century, a century that is fundamentally different from the one we still assume we live in..."

"What kind of new century are we in, then?" Reingold asked.

Drucker responded by explaining that as we enter the twenty-first century, we will be in a post-business world in which most people are part of the "knowledge" society. While in past generations no one worried about how children did on international comparative math assessments, this performance is now central to our academic concerns. "The greatest changes in our society," he concludes, "are going to be in education."[17]

Drucker sensed, I suspect, that the changes in education he anticipated would largely be in demand by employers for greater and more diverse preparation for employment. The same Spellings Commission report that announced our declining global position in educational achievement noted that "Ninety percent of the fastest-growing jobs in the new knowledge-driven economy will require some postsecondary education"[18]—in a country in which 30 percent of our students are not even getting through high school with their class.

Thomas Tierney noted in his coauthored report on the state of higher education in the United States that, of every 100 high school ninth graders, only 68 were graduating with their class in 2006. Of these, only 40 went directly on to college, and, by the beginning of the second year, only 27 were still enrolled. Of this group, 18 completed an associate degree within three years, or a bachelor's degree within six.[19] Granted, some of those who drop out of high school or choose not to enter college immediately after graduation do eventually continue their educations, but the facts are still undeniable: far too few of our young people are finishing school with the skills needed to compete in the new global marketplace.

Those who do enter postsecondary education show an alarming degree of underpreparation for college work. Forty percent of all new college freshman require remediation in reading, English communication, or mathematics.[20] About half of our undergraduates are now enrolled in community colleges,[21] and 60 percent of this group requires remediation in one or more of these critical skill areas.[22] And if they need developmental assistance when entering college, the failure rates are staggering.

A presentation prepared for the Texas Commissioner of Higher Education in 2008 by Byron McClenney of the University of Texas at Austin's Community College Leadership Program put a state-specific face on what are often featureless numbers. The report indicated that, in Texas, 66 percent of the students entering community colleges between 2003 and 2006 were not "college-ready" by Texas assessment standards. Eleven percent were deficient in all three of the evaluated areas: mathematics, reading, and writing. Twenty-five percent of students admitted to the state university system during the same period were also deficient in one of these critical skills. More alarming is the fact that only three of ten deficient students entering community colleges successfully completed the developmental math sequence within *four* years of college enrollment, while only half of those who needed remedial reading or writing completed those courses successfully. In mathematics, a mere 17 percent who tested into developmental math when admitted to college in 2003 successfully completed a single college level math course by the end of the 2006 academic year. This means that of 44,745 students who tested into developmental math, only 7,694 passed a college level math course within four years![23]

Data collected from a group of 27 community colleges that participated in the inaugural year of an initiative entitled "Achieving the Dream," sponsored by the Lumina Foundation, revealed that, of every 100 students who entered these colleges with a need to take remedial math, only 8 completed a college level math course within three years.[24] And all of this is occurring in a country where 90 percent of the fastest growing occupations require some college preparation.

The "perfect storm" in American education that ETS wrote about is based on the idea that three critical factors are converging at this moment in our nation's history. The first two have been mentioned—a growing skills gap between what the workforce needs and our level of preparation, and changes in the world's economy in ways that require higher levels of education and greater global awareness. The third is that our national demographics are shifting in such a way that those retiring from our workforce cannot be replaced by individuals with the same levels of education and training—at a time when *better* education is needed to remain competitive.

At the middle of the last century there were seven people working to help support the retirement needs of every elderly person in the United States. By 2030, that number is projected to be three.[25] Between 2006 and 2016, the 55-and-older age group in the American workforce is expected to increase by 46.7 percent, while the 25–55-year-old group will grow by only 2.4 percent, and the numbers entering the labor market from the 16–24-year-old group will actually decline by nearly 7 percent.[26] Major companies, even in the United States, will be required by necessity to look elsewhere for labor—particularly if this declining number of new entries is underprepared.

Several years ago during a flight to China I was seated next to a representative from one of the major American software manufacturers, and during the course of our conversation I asked what he thought about outsourcing labor to countries such as China.

"We really don't use that term anymore," he said. "It assumes that a company or product has a home country and that sending work anywhere else is 'outsourcing.' But the designers and engineers who work for us are everywhere. Even our management is scattered around the world. So are we an American company, an Indian company, or a Chinese company? And are we outsourcing when we send work there, or to the U.S.? Do you think Toyota employees in Kentucky think of themselves as 'outsourced' labor? For most multinational companies, it's really a term that doesn't make sense anymore." The perfect storm is brewing because multinational corporations think in these terms, and other countries are becoming better educated than we are, have lower wage scales, and work can be moved by these companies to places where people are better prepared to do it, at lower cost.

Anticipating this collision of circumstances, the New Commission on the Skills of the American Workforce wrote in its *Tough Choices or Tough Times* report:

> If we continue on our current course, and the number of nations outpacing us in the education race continues to grow at its current rate, the American standard of living will steadily fall relative to those nations, rich and poor, that are doing a better job.... Although it is possible to construct a scenario for improving our standard of living, the clear and present danger is that it will fall for most Americans.[27]

A Grassroots Revolution

Even without the crippling recession of the first decade of this century, this dire prediction is coming to pass. We appear as a country to be facing two probabilities: that the current generation of college-aged students will experience a lower average standard of living than that of their parents, and that they may well be the generation that sees the United States cease to be the dominant economy in the world. The *Tough Choices* report suggests, however, "that it is possible to construct a scenario for improving our standard of living."[28] This book is about one of those scenarios. It is based on the belief that the positive change that is needed depends on our ability as a country to regain our international competitiveness in education and job preparation, and reassert our position as the world's leader in educational achievement. It argues that to regain this status we must not only have educational reform but *revolutionary* reform at all levels of education, and that it can, and probably must, be initiated and implemented at the local school district level.

I firmly believe that, despite the best intentions and huge infusion of money coming from Washington, this reform will not, and cannot, be led or dictated by federal mandate or by state government. It will not come from universities or be initiated by state departments of elementary and secondary education. Both are in their own ways entrenched bureaucracies, with universities often acting as if they are Congress with tenure—exhibiting deeply polarized certainty about almost everything, but showing very little understanding of what is happening in the world outside their own beltways. Reform must begin in communities—in small towns and midsized cities, in suburban bedroom communities, and in large urban districts. At the levels and in the places where we can *directly* influence what is happening in our schools, we must find the means to restore rigor and introduce and stimulate change. This will require a catalytic group of bold citizens who are willing to lead campaigns to get the right people on school boards, demand accountability from them, and support them as they make the very difficult decisions required to lead reform. It will demand new or redirected educational leadership at the school district level—superintendents and principals who have the courage to implement controversial policies that lengthen the school day and school year, focus curriculum on essential courses and programs, and insist on compliance with strict standards of performance and behavior. It will require like-minded communities to come together with similar groups from other cities and states to demand state legislation that allows school districts the flexibility to restructure elementary and secondary education. There is little question that it can be done. The question is whether we have the national will and discipline to do it.

CHAPTER TWO

Where Are We Doing It Right?

"Just as there are no shortcuts, there are no excuses. Students are expected to achieve a level of academic performance that will enable them to succeed in the nation's best high schools and colleges."
—"Pillar 5" of the KIPP School's Five Pillars

After the opening chapter, you may be feeling some of the cynicism and despair about education in the United States that has haunted me for the past twenty years. We are failing miserably with our current approaches to elementary and secondary education in America, and if I didn't do an adequate job of raising your level of concern, I recommend reading the *Tough Choices or Tough Times* report by the New Commission on the Skills of the American Workforce, or ETS's *America's Perfect Storm*. What these well-documented reports have to say should shake you to the bones.

Kay McClenney, who directs the Community College Survey of Student Engagement at the University of Texas at Austin, has a saying she calls one of the fundamental laws of the universe. It goes something like this: "All of the assignments, courses, and programs we now have in place in schools are designed to produce exactly the results they are now producing." A corollary to this fundamental law is that if we want different results, something has to change—and if we want dramatically different results, things have to change dramatically. We need a revolution in America's schools and, like our great revolution for independence, it needs to be led by common citizens taking a stand in one community after another, in communities across the nation.

Fortunately, there is no need to start from scratch to determine what this revolution should be like, how it can be done, or where it should lead us. A number of groups have already charted a path. As early as the 1980s, educational reformers told us that we needed more rigorous school curricula, longer school days, an extended school year, and greater emphasis on English, science, mathematics, and foreign language. They told us that faculty should be better prepared and have better pay, and that we needed more teachers in math and science.[1]

In 2005, the Association of American Colleges and Universities organized a National Leadership Council for Liberal Education and America's Promise, called the LEAP initiative. The Council consisted of college and university presidents, leaders of business, industry and labor, chief executives of prominent foundations and not-for-profit organizations, and representatives from the public media. Its charge was to recommend the essential learning outcomes required of a college education that would prepare students to meet the challenges of the new globally integrated century.[2] Through what the study identified as a multiyear dialogue with hundreds of colleges and universities, analysis of recommendations from the business community, and review of the requirements and expectations of numerous educational accrediting agencies, the Council developed a list of four essential learning outcomes it believed would prepare students for twenty-first-century citizenship, leadership, and scholarship. These included:

Knowledge of Human Cultures and the Physical and Natural World
- Thorough study in the sciences and mathematics, social sciences, humanities, histories, languages, and the arts.
  Focused by engagement with big questions, both contemporary and enduring.

Intellectual and Practical Skills, including
- Inquiry and analysis
- Critical and creative thinking
- Written and oral communication
- Quantitative literacy
- Information literacy
- Teamwork and problem solving
  Practiced extensively across the curriculum, in the context of progressively more challenging problems, projects, and standards for performance.

Personal and Social Responsibility, including
- Civic knowledge and engagement—local and global
- Intercultural knowledge and competence
- Ethical reasoning and action
- Foundations and skills for lifelong learning
  Anchored through active involvement with diverse communities and real-world challenges.

Integrative Learning, including
- Synthesis and advanced accomplishment across general and specialized studies
  Demonstrated through the application of knowledge, skills, and responsibilities to new settings and complex problems.[3]

Granted, these were recommendations for what a *college* education should provide. But the greatest challenge to producing the college graduates we need is to first develop students who come out of high school prepared for the rigors of this college curriculum—who have spent their

K-12 years in schools with similar learning objectives and similar rigor. Fortunately, there are schools across the country that meet these standards with impressive results, and they are showing up in some of the least expected places.

Discovering Excellence

Helena-West Helena, Arkansas, sits on the Mississippi River an hour's drive south of Memphis, and immediately across the river from the northwest corner of Mississippi. It is a community that has been battered by hard economic times and the loss of major industry, with the largest remaining commercial employer, the Isle of Capri Casino, in neighboring Mississippi a few miles away. Median household income in the area was $15,000 below the Arkansas average of $38,134 in 2007,[4] and the population in and around Helena-West Helena has been in decline for over a decade. Unemployment fluctuated between 6 percent and 11 percent even before the recession of 2008.[5] Helena's partially boarded-up main street is separated from the Mississippi River by a high earthen levee that springs to life once a year when the popular Arkansas Blues and Heritage Festival comes to town, attracting some of the best blues artists in the world. It is noteworthy that the festival was once called the King Biscuit Blues Festival, named after Helena-produced King Biscuit Flour. The King Biscuit Flour Hour emanated from local KFFA and was the country's longest running Blues program on radio. But King Biscuit is now gone, and KFFA sold the rights to the name of the Flour Hour to King Biscuit Entertainment in Memphis.

Yet, despite these losses, this town of 12,500 people boasts one of the more successful middle schools in the United States—a charter school run by the KIPP organization. The KIPP Delta College Preparatory School enrolls 500 students in grades K-1 and five through twelve and is adding grades each year. It is free to any student in the Helena-West Helena School District who wishes to attend and is selected by lottery—but not without a number of strictly enforced provisions. The school day is longer, running from 7:30 a.m. to 4:00 p.m., with classes or mandatory activities every other Saturday. The school year is also longer. Students must come to school in uniform, and every parent signs a letter of commitment to support the student's academic efforts. Since more students apply to Delta College Prep than the school can accommodate, a random lottery selects those who will attend, and special education services are made available to students with disabilities. The school has been so successful that the Delta Bridge Project, the strategic planning group for Phillips County, Arkansas, where the school is located, established as a strategic goal to "Expand the KIPP Delta College Preparatory School to educate children from kindergarten through twelfth grade." The Delta Bridge Project's 2007 report announced that the community had assisted in obtaining the financing and property needed to meet this goal.

KIPP schools, like Delta College Prep, focus on five basic pillars: high expectations, choice and commitment, more time committed to education, power to lead, and results. And the results produced by Delta College Prep are astonishing for a depressed rural community with a largely minority student body. On the Arkansas 2006–2007 Benchmark End of Course Exam in Algebra I, 93 percent of the KIPP school students scored at the proficient level, compared to 14 percent of students in the Helena-West Helena School District, and 61 percent for the state as a whole. On the Geometry exam, 95 percent of KIPP students scored at the proficient level, compared to 36 percent for the local school district, and 59 percent for the entire state.[6] What is particularly noteworthy is that 95 percent of the Delta College Preparatory School's students are African-American, and the school maintains an average daily attendance of 97 percent. Granted, there is a refining process that takes place as families choose to apply to the school, agree to comply with the strict standards, and insist that their children work hard and achieve. But these are all refinements created by *family choice*, not by institutional selection or exclusion. Since Delta College Prep must select by lottery from those who apply, it essentially ends up with a student body that is a random representation of the families who want this opportunity for their children. (Some in Helena argue that the selection process isn't truly random, and that KIPP "recruits" students from the local schools who are seen as high achievers. But the KIPP administration insists that the lottery system is used to draw from all applicants, and that family choice selection is the greatest factor in refining the student body.)

Since Delta College Prep students enter at the fifth grade level, they have been in the regular public school through grade four. State performance data show steady, upward improvement from the year they enter. As the 2008 state performance data for the school in the table below indicate, students demonstrate steady progress as they move from grades five through eight.

Across the nation, KIPP's seven high schools and 52 middle schools have enrollments that are 90 percent African-American or Latino, and more than 80 percent of these students go on to college. At the high school level the curriculum requires four years of English, four years of mathematics, four years of science, four years of social sciences, two years of humanities, and two years of world languages.[7] The school year, school day, curriculum, and academic rigor look remarkably like the recommendations of the *A Nation at Risk* study of thirty-five years ago, and are producing just the results the Commission hoped for.

In Chicago, the Noble Street Charter schools are achieving similar results with a similar student body and similar policies. Though also nonselective, the schools require parents to attend four parent/student/teacher conferences per year, read and sign a weekly newsletter, and sign a biweekly student progress report. Like KIPP, Noble requires a longer school year, longer school day, and school uniforms. Average daily attendance for the year is 95 percent, and 80 percent of freshmen who enter the system graduate from high school, with nearly 95 percent of graduating seniors going on to college.[8] Noble now operates seven high schools in the Chicago area, surrounded by some of the poorest performing high schools in the nation.[9] Similar results have been attained by other schools across the country, also located in some of the nation's most troubled districts.

Freedom to Raise Expectations

The lessons seem clear. When schools, communities, and parents seek excellent education for their children and are willing to support the schedules, expectations, and academic rigor required to produce excellent results, students rise to the occasion. In both the Noble model and the KIPP model, each individual school has some discretion about how the curriculum is structured, but the fundamentals are similar: four years of math, science, English, and social sciences, with additional coursework in the humanities and world languages. Electives are available, but only after this core schedule has been met.

From the first day of class, these schools stress with students that academic achievement is what the school is all about. At Delta College Prep, new students have to "earn their desk," and sit around the outside of the room on chairs during the first few class sessions, until they demonstrate to their teachers that they are serious about assignments and homework and have come to achieve. Participation in extracurricular activities—of which there are many—is dependent on academic performance, and there is none of this "we understand that if you are going to play basketball, you will have to be excused from some of your academic work." The understanding by all students is that "if you are going to play basketball, you must first meet all of your academic expectations."

KIPP and similar charter models are not without their detractors. Some data show that weaker students drop out and return to other schools in the districts, inflating performance results for the charter school over time. Others view the policies of charter schools such as KIPP's Delta College Prep as bordering on the draconian, forcing a discipline and set of expectations that stifle creativity and growth, while providing students from minority backgrounds with false hope that they can succeed in the mainstream. Jim Horn in his essay review of *Washington Post* writer Jay Mathews' laudatory book about KIPP refers to the KIPP environment as "hostile to the well-being of children," and notes that the system's acronym is referred to by some as standing for "Kids in Prison Programs."[10] Still others believe that charter schools are too free of accountability and oversight, while robbing school districts of critically needed resources for day-to-day operations of other schools. My experience with KIPP schools does not support these concerns, but for the sake of these critics I should explain now that this is not a treatise in support of charter schools.

I acknowledge, in fact, that some of the schools listed as being among the worst in the country—including the one mentioned earlier with the graduation rate below 10 percent—are also chartered. The Center for Research on Education Outcomes (CREDO) at Stanford University published a comprehensive report in 2009 that demonstrated that, in the aggregate, students attending charter schools did *less* well than those attending public peer institutions in the same communities. The same report notes, however, that students learning English as a second language and those from economically disadvantaged backgrounds who attended charter schools did appreciably better.

It is also important to note that another study conducted in New York City the same year as the CREDO review found that charter schools did considerably better than the Stanford report had acknowledged, when the research was corrected for what the New York study called "switcher" bias. Expressed very simply, the three researchers conducting the New York City Charter Schools Evaluation Project argued that when a study compares students who switch to a charter school after starting in the public schools to students with the same characteristics who did not choose to switch, a natural bias is introduced. This bias results from the study's failure to take into account the reason the student's family chose to move the student—a reason that may very well be a reflection of signs that the "switched" student was beginning to struggle in the previous school environment. In order to remove "switcher" bias, the only statistically defensible way to compare student performance, according to the New York study, is to look only at students who asked to be part of the lottery draw to be admitted into the charter schools, then compare performance of those with similar backgrounds and academic achievement up to the point of the lottery who were selected-in, or selected-out.[11] The New York study found the lottery process to be unbiased, and the two selected-in and selected-out pools to be essentially identical. But when the performance of the students was compared in subsequent years, the selected-in students did significantly better on the New York Regents assessment of basic skills test, scoring an average of three points higher for every grade spent at the charter school than did the selected-out students. They were also 7 percent more likely to complete a Regents diploma by age twenty.[12] The New York study indicated that keys to greater student success were a longer school year, more minutes devoted to English instruction during each school day, a discipline code that expected small courtesies and punished small infractions in the classroom, teacher pay based partly on performance, and a mission statement that emphasized academic performance as the primary goal.[13]

Even in the CREDO study where switcher bias was not accounted for, in some states and given certain chartering freedoms and limitations, charters outperformed the public schools by significant margins—with Arkansas serving as a case in point. This suggests that there are important reform lessons to be learned from those who have demonstrated extraordinary performance results that the nation must consider, and that we should seek to provide these opportunities to *all* schools.

Applying the Charter Model

The secret to successful reform lies *not* in the freedom that is given to charter schools to operate differently and change policies without unnecessary interference, but in their freedom to use that flexibility to enforce standards, and send students elsewhere to learn who don't wish to comply. It is true that, in some ways, charters extend to schools a flexibility that frees them from oversight—allowing those who choose to exercise these freedoms irresponsibly, or without a clear plan to raise standards, to fail miserably. This, I would argue, is a failure of oversight requirements, not of extended freedoms. Schools that use flexibility to raise academic expectations are providing some of our nation's best examples of unusual success with challenged student populations. Those that elect to experiment with untested curricula, allow students to chart their own academic paths, or immerse themselves in "identity studies" are often much less successful at strengthening academic performance and at meeting what are supposed to be mandatory reporting and improvement expectations if chartering rights are to be retained.

Ben Chavis who directs a successful charter school in Oakland, California, believes that these schools fail because they do not use the freedoms they have been extended to "break the status quo public school mold."[14] Linda Mikels who served as principal of Sixth Street Prep in Victorville, east of Los Angeles, attributed the failures to experimenting with unproven instructional methods, and forgetting that student success was their primary responsibility.[15] Other successful school leaders have explained the failures in terms of poor fiscal management and an unwillingness to expect students from low income or minority backgrounds to perform up to the standards expected of others. All agree, however, that there is a list of expectations that can create success and that schools with the freedom to pursue these expectations can succeed with students from any background. Chester Finn, Brunno Manno, and Gregg Vanourek note in their comprehensive evaluation of the charter school movement, *Charter Schools in Action:*

> The genius of the charter concept is that it is demanding with respect to results, but relaxed about how these results are produced; tight as to ends, loose as to means. Yet success in attaining results can be found only if there are clear standards, good assessments, and consequences to everyone.[16]

One might argue, in fact, that if the Stanford CREDO report is accurate and only 17 percent of charter schools are performing at levels that

significantly exceed that of peer institutions, states have not been demanding with respect to results or tight as to ends, and should have pulled
the charters of those schools that have been allowed to continue without
showing success. But if we are to bring broad-based school reform to
education in America, schools in every community must be given this
same freedom to innovate, modify, hire, fire, and enforce standards, while
being held accountable for their performance.

Forty of the fifty states and the District of Columbia now have laws
that provide for the creation of charter schools. The movement began
in the early 1990s when Minnesota passed the first charter school law in
1991, followed by California the following year.[17] Although charter laws
differ from state to state, they generally have seven areas of commonality that describe who may propose and grant charters, how the school
is legally defined, how funding will be provided, how admissions will
be governed, staffing and labor issues, control over instruction, and how
accountability will be measured.[18] The Center for Educational Reform, a
nonprofit advocacy group for the charter movement based in Washington
DC, has ranked state charter school laws and considers the Minnesota law,
amended in 2006, to be the strongest. Characteristics that appear to make
these laws particularly effective include:

- No limit is placed on the number of charters that can be issued in the
  state. The CREDO report (which looked at only fifteen states and
  the District of Columbia) indicated that states with caps on the number of charters that could be issued had lower student success rates.
  One speculation as to why this might be the case is that organizations
  that have been successful in running high performing charter schools
  are more likely to go to states with broader chartering opportunity.
  Another possibility is that, in states with limitations on the number
  of charters, the few that are issued go to schools that are designed as
  "alternative" choices for students who for disciplinary reasons, preg-
  nancy, truancy, or other reasons are unable to stay in the traditional
  school. In my own state, Missouri, charters are currently limited to
  communities with populations over three hundred and fifty thousand, meaning that only urban areas with the most serious school
  challenges can develop charter alternatives.
- Local school boards, colleges and universities, or nonprofit organizations can apply for charters, with the approval of local school boards
  and of the state Commissioner of Education. The CREDO Report
  indicates that limiting who can *approve* charters is important, and that
  states that allow more groups or organizations to approve charters do
  less well. The implication here might be that approving authorities
  other than the state's department of education are less concerned with
  insuring that academic performance is central to the chartering plan.
- Applications denied by local school boards can be appealed to the
  state commissioner. States that have an appeals process for schools

whose charter application has been denied appear to have stronger charter schools.

- Most state and district education laws and regulations are waived for the charter school.
- There is broad legal autonomy.
- Governance requirements do not require the standard administrative certifications that are common in most states for school leadership positions.
- Transportation for students is provided by the local school district, and there is some assistance with facilities.
- There are strong state reporting and accountability requirements.
- State per-pupil funding follows the students.
- Teacher certification may be waived if other credentials qualify the teacher.
- Admission requirements are not permitted, and schools must use a lottery system for selecting students from those interested in attending. The racial balance must reflect the region being served.
- Schools must include performance expectations in the chartering agreement, and must demonstrate results in order to maintain the charter over time.

The most immediate and effective way to transform education in America is to create model charter laws in each of the fifty states, and extend these chartering opportunities to entire school districts. Any community within a state must be given the opportunity to "charter" its school district, removing it from the limitations imposed by state laws and regulations and allowing it the freedom to hire the leadership it desires, extend the school day and school year, revise curriculum without laborious state approval processes, and select and create expectations for faculty. The deciding criterion for maintaining the charter must be student performance—not arbitrary or historically established policy and procedure. If the district is demonstrating that students are achieving at or above an acceptable level to the state, the charter is maintained. If performance falls below those levels, the charter is revoked and the district returns to the operating conditions previously applied by state statute. Without this opportunity for community-based reform, the restructuring of education we need so critically in the United States will not happen, and we will continue to fall further behind.

A number of the characteristics of successful charter laws may appear to be teachers' union's worst nightmares: alternative teacher qualification and certification, freedom to make dramatic revisions in curricula, strict teacher accountability, broad legal autonomy, and freedom from many of the state's laws and regulations governing education. But schools given this latitude are performing remarkably well with student populations that have traditionally been some of the least successful. In communities such as Oakland, California, where several successful charter schools operate, some faculty have remained union affiliated and others have opted out. The lesson seems to be that control of these factors must be returned to school boards who exercise their new authority with one objective in mind: to increase student performance. Faculty who are comfortable with contracts that focus on measured accountability find the learning environment very much to their liking. It is important to state again that it is not "chartering" that creates success, but the latitude to introduce reform-directed change without the restrictions that over the years have become the bureaucratic sludge that clogs the engine of state departments of elementary and secondary education and impedes school reform.

There are, unfortunately, two major limitations to extending charter-like opportunities to every public school district. The first is that, as noted earlier, a number of states do not have chartering provisions at all, and so seem unlikely to embrace changes in statutes that would extend these freedoms to their general school populations. Many states that do allow charters have statutory limitations on the number—so, statewide expansion of the flexibilities offered by the charters will require aggressive public engagement.

In addition, charter schools by their very nature have one luxury that sets them apart from districts without this option—a luxury that is as critical to their success as are longer school days, a lengthened year, or revised curriculum. They have the choice to deny admission or continued enrollment to a student or family that refuses to agree to follow the prescribed standards. In fact, *choice* is at the heart of charter school success, and is fundamental to school reform in America.

Yet, *choice* is a term that has become loaded with political connotation and, in some people's minds, smacks of vouchers, reallocation of public dollars, and shuffling students between districts. When the issue of choice has been raised in the school quality debate in the past, it has generally been in reference to students and families having the choice to go elsewhere if the school is not performing. More important to educational reform in the future is extension of choice to schools to move students into alternative learning environments *within the public district* if the student chooses not to comply with expected standards of performance and behavior. The system must be able to enforce a level of discipline and academic expectation that creates the standards of performance the community expects and requires, while providing those who do not wish to meet these expectations another learning option.

It is not accidental that one of the Five Pillars of the KIPP school network is *Choice and Commitment*. It is an expression of the philosophy that if students and their parents do not wish to comply with KIPP standards, they can attend elsewhere. Under this philosophic "Pillar" the Delta Prep School explains:

> Students, their parents, and the faculty of each KIPP School *choose* to participate in the program. No one is assigned or forced to attend

our school. DCPS is an open-enrollment charter school that does not choose the best students in the community to achieve success; in fact, incoming fifth graders are admitted regardless of their test scores. The only requirement for admission is the willingness of each student and parent to sign and uphold the *KIPP Commitment to Excellence Form*. Teachers also sign this form so that all parties are held accountable for doing whatever it takes for the education of the student.[19]

For a school district charged with serving all students, this is immediately problematic. When I asked the superintendent of one of the KIPP schools what he would do if there weren't a public school in the district to which unmotivated and unwilling students could be directed, he said, "There is always choice. If a district decides to have standards and expectations like we have here, parents can decide if they want to live in this district or elsewhere."

Obviously, it isn't that simple. For reasons of employment, family connections, or just geographic preference, families will choose to live in specific communities. The school can't just say, "If you don't like what we require, move." But the school *can* say, "This is what we require. We have created an alternative for those who don't wish to rise to our standards, but if you participate in the standard school curriculum in this community, you will be expected to meet our rigorous expectations." (I talk more about creating an alternative choice in chapter 12.)

Critics will argue that you can't simply let every community decide how long its school year will be, what curriculum it will provide, who it will select to teach its children, and who it views as capable of managing its school system. But the truth is that yes, you can. Chartering has shown us that all of these opportunities can be extended to a school and can produce exactly the results we want from public education. In fact, this community-based choice is the only way excellence will be achieved.

In truth, many of the factors that are leading to improved student success in our best schools—rigorous curriculum, lengthened days, and longer school years—can be implemented without chartering approval. Most states impose minimum requirements in these areas, but districts can choose to do more. Community-based reform can begin in any community at any time, by determining what students must learn to remain competitive and college ready, and by making the adjustments needed to move in that direction. As Terry Suarez, a friend and colleague in education in western Virginia, reminds me each time I see him, "We need to create a 'culture of education' in each community in America."

Achievement Without A Charter

When I returned from my visit to Arkansas' Delta College Prep, I commented to my research assistant, Mirra Anson, that *surely* there must be successful public schools outside of the charter movement that are following principles that help high performing charters work so well.

"You need to go to Leopold," she said. Before joining our department for doctoral work, Mirra worked as an admissions representative for another university in the southeast corner of the state, and she had been assigned to visit this small community. Her colleagues in the Admissions Office assured her that this would be one of the highlights of her recruiting stops, and her first visit demonstrated why. She directed me to the Leopold School District's web page where the headline announced "The Way Education Should Be."

Between 2000 and 2009, Leopold R-3 was awarded the "Distinction in Performance Award" by the Missouri Department of Elementary and Secondary Education every year, and for the 2006–2007 year was chosen as the best school in the southeast part of the state by the Southeast Missourian Newspaper. The following year, the high school was honored at the state capital with the Gold Star Award, recognizing that its students had performed in the top 10 percent of all schools in Missouri on the state assessment exams—no small feat for a school that meets in a collection of metal frame buildings in a tiny rural town in the Missouri Bootheel. Seventy-nine percent of its students tested above the national average on the ACT exam in 2007 (the common assessment in the central Midwest), and average daily attendance was 97 percent. No senior dropped out of school that year, and 86 percent of graduates went on to a two- or four-year college. I decided I needed to visit Leopold.

The first trick was getting there. Leaving Interstate 55 at Perryville, the drive south on Highway 51 leads through Marble Hill, where a turn onto Highway 34 takes you to Country Road 408, then 403, and on into Leopold. The school and church are together in the middle of town and are, in every way, centerpieces of the community. Superintendent Derek Urhahn tells you with some pride as you tour the three simple classroom buildings that, in ten of the eleven years he has served as superintendent, no parent missed a parent-teacher conference for an elementary student— and only one missed during the eleventh year. "Something came up that couldn't be avoided," he says.

I told Superintendent Urhahn that I had driven through a dozen little towns just like Leopold on my way down and knew that in terms of student achievement their schools didn't hold a candle to Leopold. What made his school and community different? He suggested I ask Patty Bohnsack, the district's secretary/bookkeeper, who had grown up in Leopold and completed her education there.

"It's been this way for generations," she said. "We're a town of old German immigrants and started with a Catholic school there by the church. When we decided we needed a public school, that building was leased from the church and people wanted the same emphasis on education. It's just always been part of the culture."

In the small library, assistant Tammy Vandeven, also a Leopold graduate, said that every child starts school expecting to do well. "From the day you walk in, they push you," she said.

"I prefer the word 'challenge,'" said Bobby Jansen, who is a third generation Jansen on the Leopold school board. "From before the time I was a student here, our philosophy has been that every student needs to be challenged." It's a sentiment that is clear in the board's statement of philosophy that "In carrying out its responsibilities, the Board of Education is guided by the desire to use the resources of its community, its staff and its students to provide the highest quality education permitted by its financial resources. In reaching decisions the Board will attempt in every case to act in the best interests of its students."[20]

When asked how the board carries out this philosophy, Jansen nodded toward Superintendent Urhahn and Principal Keenan Kinder and said, "We don't just challenge our students, we challenge our teachers and administrators. They know our standards, and we all have a role in meeting them."

Second-year teacher Shawn Kinder says that a new teacher quickly learns that students are expected to achieve and teachers are expected to help them do it. "Your colleagues let you know right away that we expect our students to do well," he said. "But more importantly, the students let you know that they expect to do well, and want you to help them do it. I taught for five years before I came here, and I've never seen the kind of total support you get here. Board, administration, community, students. It all comes together."

Next year, senior Kelli Woodfin plans to begin her premed studies at Southeast Missouri State in nearby Cape Girardeau. Does she feel in competition with her classmates?

"Sure, but in a friendly way," she said. "We all want to do well and there's a lot of competition for the valedictorian and salutatorian. But we want each other to do well too."[21]

Leopold exemplifies in a non-charter school many of the attributes that create a globally competitive program. Strong board and parental commitment to student success challenge students and teachers alike to stretch themselves, and both are evaluated on performance. In the community, there is a "culture of education." After visiting the school I again asked Mirra what had impressed her about Leopold to inspire her to recommend it as a model. "I think it was the sense of trust I saw there," she said. "The community trusts the school to do what is in the best interest of the students. The students trust their teachers, and the teachers trust their administrators and board. As their motto says, it's 'The way education should be.'"[22]

To many, this will sound like Shangri-La—a small remote community untouched by the brutal realities of urban sprawl, clashes of cultures, and the challenges of meeting the needs of disadvantaged populations. But

Leopold is not alone as a non-charter example of academic excellence. In El Paso, Texas, Ysleta Independent School District (ISD), a huge, largely Hispanic district, has made the same commitment to excellence using many of the same principles and is succeeding in the face of some of the most challenging circumstances. I will discuss Ysleta at greater length in chapter 4. These successes are not reflections of location or of resources, of school size or of state support, but of community commitment and the decision to become great schools.

But the broad *revolutionary* reform we need will necessitate sweeping changes in leadership, teacher expectations, and freedom to reassign students who choose not to comply with academic and behavioral requirements—changes that can create that culture of education so often missing. Bill and Melinda Gates observed in their annual Foundation letter to the public:

> Many of the small schools that we invested in did not improve students' achievement in any significant way. These tended to be the schools that did not take radical steps to change the culture, such as allowing the principal to pick the team of teachers or change the curriculum. We had less success trying to change an existing school than helping to create a new school.[23]

For the same political reasons that reform is unlikely to occur at the state and federal levels, it is improbable that states will grant permission for all districts to "start again from scratch." But it is feasible for the best in chartering provisions to be extended to all public schools, allowing leaders to take the essential steps necessary to dramatically restructure and begin to shape this culture.

CHAPTER THREE
_____

Where Reform Won't Happen

"All Politics is local"
    —Tip O'Neill

As a former college president, I spent countless hours in the state capital meeting with legislators, talking to lobbyists, giving testimony, and attending committee meetings as special-interest groups presented their cases and legislation was debated. In one memorable legislative year, the community college presidents in the state were faced with a particularly distasteful bill to create a new technical college from a small secondary vocational school that served several rural counties in a sparsely populated region. The bill was sponsored by a powerful senator from the school's hometown who had pledged to his constituents that he would turn their little vocational school into a college with a statewide mission focused on selective high-tech career programs.

The college presidents knew that there was no need for a statewide technical college, particularly one located in this remote, rural part of the state. They also knew that if the new institution were created, it probably would not offer anything particularly novel in the way of high-tech programming, but would draw revenue from the budgets of the existing college districts. It was, by everyone's assessment but the Senator's and his rural supporters, a boondoggle display of personal power, and a bad idea.

The problem was that this senator chaired the powerful Appropriations Committee through which several funding bills important to the community colleges had to pass—bills that impacted workforce development, technology improvement, and facility maintenance and repair. The senator couldn't single-handedly stop the general appropriation to the state's colleges, but he could kill any or all of these special funding bills by simply refusing to put them on the calendar to be heard by his committee, or by keeping them low enough on the calendar that they didn't surface for debate in the time allotted for the session. And he promised the college presidents that he would do just that if they opposed the creation of his new college. I am ashamed to say that, as presidents, we caved in—justifying our poor judgment and weak spines with some utilitarian argument about "the greater common good" and extending educational opportunities to other parts of the state.

From that lesson and many like it, I have developed my own personal list of Fundamental Laws of the Universe as they relate to politics and legislation. The first is that former U.S. House Speaker Tip O'Neill was right. In a democratic system where elected officials have a constituency, "all politics is local." Politicians live by their own two R's—Recognition and Reelection—and both largely depend upon keeping the voters satisfied at home. It is the rare representative who voluntarily chooses to step down after a sensible time in office. There is something seductive about power and position, and being reelected becomes a consuming preoccupation. Once in office, elected officials work hard to keep their constituents happy, while seeking opportunities to climb the political ladder and support projects that will memorialize them.

Other fundamental laws of political life include:

- It is much easier to kill legislation than to pass it.
- Much of politics in a democratic society is horse trading and compromise, and compromise does not encourage radical reform or movement toward excellence in performance.
- Politicians generally prefer to "let the people decide" about truly controversial issues, rather than risk alienating a portion of their constituency.
- There are always at least two sides to every issue—usually more—with surprisingly little common ground in between.

Assuming that these fundamental assumptions are true—or even close to true—we should not expect to see serious reform in education come from Washington or from state government. It is against the nature of the beast. Despite the significant investment the Obama administration has made in touting education reform, and the rhetoric that is coming out of the Nation's capital concerning the need for better performance by our schools, the changes that need to be made will not happen if left in the hands of federal and state governments. As President Obama said in a national address on healthcare reform, in Washington "the default position is inertia."[1]

The Problem with Federal Reform

When we consider some of the other truly daunting challenges we face in our society—healthcare reform, social security reform, environmental reform—all issues that directly affect us in visible, tangible, or monetary ways—we get a better sense for how difficult it is to bring about major change through state and national legislation. These problems have plagued us for decades, and each administration balks at confronting them in substantive ways or finds that the process gets in the way of more than small, incremental change. Imagine how much more difficult it is to bring about broad-based reform in an area that many of us don't view as touching our lives in the personal ways we are effected by healthcare and social security. Machiavelli observed that "There is nothing more difficult to take in hand, more perilous to conduct, or more uncertain in its success, than to take the lead in the introduction of a new order of things because the innovator has for enemies all those who have done well under the old conditions, and lukewarm defenders in those who may do well under the new."[2]

A segment on NPR's Diane Rehm Show in June of 2009 illustrated the problem. On this Friday News Roundup Hour, Diane's guests were Susan Page, Washington Bureau Chief for *USA Today*, Eamon Javers, a correspondent with *Politico*, and Laura Meckler, reporter for the *Wall Street Journal*—all keen Washington observers.

A caller asked about all the difficulties that were plaguing the Obama Administration as it attempted to push its healthcare agenda forward, and Diane Rehm wondered aloud if people marching in the streets to insist on better, less expensive, and more uniform healthcare might force Washington's hand. All of the panelists immediately said in unison "It won't happen," to which Laura Meckler added, "You get that on protesting the war maybe...but anything short of that...you know, it doesn't happen." She added, "The country is just not prepared in general to do huge changes. They're much more likely to do incremental changes."

When a follow-up caller asked about the tendency toward incrementalism in American government, Meckler observed that incrementalism is the outcome of political disagreement and must be expected when people feel passionately about both (or several) sides of a hotly contested issue. The result, at best, is compromise and very small increments of progress.

A moment later, Eamon Javers added, "There are no incrementalists. There are only absolutists who run into other absolutists."[3] The basics of education reform, how it should happen, and who should lead it are issues about which people disagree—absolutely.

Using this same healthcare debate as fodder for his eternal pessimism about how Washington does business, *Rolling Stone's* often conspiratorial political affairs writer, Matt Taibbi, complained that "Just as we have a medical system that is not really designed to care for the sick, we have a government that is not equipped to fix actual crises." In an observation that applies equally to reforming education, Taibbi observed, "What our government is good at is something else entirely: effecting the appearance of action, while leaving actual reform behind in a diabolical labyrinth of ingenious legislative maneuvers."[4]

John Kingdon offers a somewhat kinder assessment of why we often see a dominance of politics over rationality, suggesting that significant policy shifts occur only when we see the convergence of three "process streams": the streams of recognized problems, policy proposals, and supportive politics.[5] With so little agreement about the nature of the problem, such divergent policy proposals, and our heavily polarized political climate, this convergence may be the equivalent of "a government that is not equipped to fix actual crises."

Roots of Federal Influence

While Taibbi's criticism may be unduly harsh, it might be helpful here to take a brief side trip into the early history of education policy development in the United States to see where the camps of absolutists referred to by Eamon Javers come from—beginning with the colonial period. By the time our Constitution was penned by the Founding Fathers, colonial school systems were well established from the elementary through college years. Harvard had existed for nearly a century and a half, and each of the colonies had its own form of college education, established largely through charters issued by colonial governments.[6] Through the first half of the seventeenth century in the American colonies, schooling of children was considered to be primarily the home-based responsibility of parents or guardians. A Massachusetts law in 1647 mandated a more formal approach to elementary schooling for communities over a certain size and was, to some degree, created to compensate for cases in which parents were derelict in their educational duties. Where schools did exist, many were church-sponsored, and schooling costs were paid by families.

When the Massachusetts colony finally introduced free public schooling in the mid-eighteenth century, support came from local rather than state taxes, and education was viewed as a community responsibility.[7] By the time the Constitution was framed in the latter half of the century, there was no sense on the part of the framers that education needed to be a federal concern, so responsibility for public education was not mentioned in this new federal charter. It therefore fell under the Tenth Amendment that states "The powers not delegated to the United States by the Constitution, nor prohibited by it to the States, are reserved to the States respectively, or to the people."[8] Education, and by inference education reform, became the responsibility of the various states or of communities within those states, setting the stage for one group of absolutists. This group continues to maintain that the federal government should not be involved in the business of education at all, and in its more conservative manifestations believes that the U.S. Department of Education should be abolished.

Yet, over the past century the federal government has gradually inserted itself into the education policy arena through its use of flow-through dollars in the form of grants and assistance to states. Pieces of legislation, such as the Elementary and Secondary Schools Act of 1965, the Child Nutrition Act of 1966, and the Carl D. Perkins Vocational and Technical Education Act of 1998 define policy, provide block grants to states for specific purposes, such as school lunch programs, then impose mandates and policy expectations on the states and school districts if they wish to receive this federal money. Since virtually all states and public schools believe they need this support to operate, they are essentially committed to follow federal guidelines that accompany these laws. Much of this federal money is directed through state departments of education that are then committed to follow federal mandates, and require state schools to do the same if they are to receive flow-through dollars. Proponents of federal involvement in education point to our history of segregation, to the inequities created by pockets of poverty that seemed to go unaddressed by state governments, and to the interventions at the national level that desegregated schools, introduced school lunch programs, and provided equal opportunity for women and minorities as justifications for this involvement. They argue that essential educational progress would not have happened without federal intervention and won't in the future—creating our second absolutist group. In the middle somewhere—or at a corner of a triangle—are those who are strong proponents of private education, but see no reason that public dollars shouldn't be made available to allow students and their families to choose how and where these dollars are spent on education; absolutist group three. Each of these groups believes reform to be necessary and each has its favorite approach, with some of their ideas miles apart.

One of the most recent federal mandates, and one that illustrates perfectly why revolutionary school reform will not come from the federal government, is *No Child Left Behind*. This reform effort included some of the elements of effective change that were discussed in the last chapter—greater emphasis on accountability, greater local flexibility, and greater choice for parents. But it lacked the essential ingredient needed to make schools better: the ability to change public school policy and curriculum in meaningful ways without strict state oversight by Departments of Education. Instead it relies on test results based on state-developed assessments to determine how schools are doing, and threatens them with sanctions unless they meet performance goals established using these measures.

The major complaints leveled against *No Child Left Behind* are that it forces schools to group all students together when assessing levels of proficiency, and that it places great pressure on teachers to teach students what the state-developed tests emphasize, rather than what children should effectively learn. The first criticism has some merit, in that schools with high numbers of students who are learning-disabled, or with large recent immigrant populations who are not native English speakers, are evaluated side-by-side with stable, affluent suburban districts with few students with special needs. Teachers in this first school (which we will call School A) complain that although they may do a marvelous job of helping their disabled students progress, or of assisting immigrant students with basic language development, they will not show levels of student proficiency or improvements in performance on state exams that are close to those achieved by students in suburban School B.

Malcolm Gladwell argues that, as a result of a long list of circumstances having to do with when and where a child is born, who her parents are, and what changes are occurring in the culture around the child, many of the students in School A are already left behind. It will take a Herculean effort by an exceptionally talented and motivated core of teachers to move this group forward.[9] This isn't to say it can't be done, but it does point out the inequity of measuring progress in classrooms in Schools A and B by a common ruler.

Granted, *No Child Left Behind* did provide School A with what is called the "Safe Harbor" provision—a caveat stating that if a school missed state progress goals for the percentage of proficient students but at the same time succeeded in reducing the number of students below the proficiency level by at least 10 percent, the school still achieved its AYP (Annual Yearly Progress) goals. But one might think about that as an invitation to "leave the challenging children behind" while focusing on the group that can most easily boost proficient rates by 10 percent.

Who Gets Our Attention?

I recall a particularly gut-wrenching session with a group of faculty and staff at a community college where the Institutional Research Director has developed an effective risk assessment model for incoming freshmen. Typical of many community colleges, 60 percent of entering students required developmental education in one of the basic academic skill areas. By plugging a series of factors such as high school grade point, ethnicity, employment workload, gender, and number of required developmental courses into his formula, the IR Director could generate a "risk index" that predicted with about 70 percent accuracy whether the student would make it into the second semester. Using his risk index data, he had also examined how various academic interventions such as tutoring, mentoring, and supplemental instruction affected the likelihood of changing these persistence rates.

The data indicated that students could be divided roughly into three groups by risk factors, from low to high. The students with the lowest risk indexes (minimal risk) were likely to persist into the second semester with or without interventions. Students in the highest risk factor group failed to persist, even with significant assistance. If the interventions were applied to the students in the middle risk category, the change effect was by far the most significant. The debate in this meeting centered around the question, "If we have limited resources that allow us to provide special assistance to only a portion of our students, who should get the extra help?" The students who are the most in need? The students who are already successful but might be even more successful in the long term if given enhanced learning opportunities? Or the group in the middle, where we will see the greatest statistical gains if we commit our resources there?

The Safe Harbor provision forces this decision in a specific direction— toward the students who will statistically show the greatest progress, but may not be in greatest need or have the greatest opportunity to excel. It encourages schools with limited resources to pull them away from "highly at risk" and "enrichment" opportunities, and direct them at the statistically responsive center. As author and legal expert Phillip Howard observes, "We seem to have achieved the worst of both worlds: a system of regulation that goes too far while it also does too little."[10]

The complaint that *No Child Left Behind* pressures teachers to teach to the state test rather than to some other measure of educational achievement is valid only if there is evidence that the state exams do not accurately measure what we commonly agree a child should know and be able to do. It is true that statewide assessment may limit teacher flexibility and latitude to teach whatever is of greatest interest to the faculty member, or what the teacher may think students should learn. But if the assessments are constructed based on the judgment and consensus of the best group of educators the state can bring together, one might assume that a well-prepared and broadly educated student would do well on the exam. The Delta College Preparatory School's success on the standardized Arkansas assessment tests supports the contention that students who are well-prepared through a strong, content-rich curriculum will do well, even if they come from backgrounds from which students often have performed poorly.

There is, however, a basic flaw in a program that provides rewards or penalties from the federal government to states that show statistical improvement or regression on state-developed and regulated assessment instruments. Diane Ravitch, who served during the first Bush Administration as Assistant Secretary of Education, has been critical of *No Child Left Behind* because of the controls states exercised over testing and reporting. "It did nothing to raise standards," she said, "because it left decisions about standards to the states. So many states have very low standards and yet announce that more and more students are 'proficient,' as defined by that state." Ravitch adds:

We know from the National Assessment of Educational Progress that actual improvement has been very small over these past seven years, and in some cases, the rate of improvement has been less in these past seven years than in the years preceding the passage of NCLB. In the meantime, schools have become test-obsessed in a way that is not conducive to good education. Many schools and districts and states have learned how to game the system, and they are producing higher scores (by lowering the passing mark—or cut score—on their tests) that do not represent genuine improvement in learning.[11]

Ravitch expresses strong opposition to the charter movement and to privatization, citing the CREDO study as an indication that Chartering programs don't produce better achievement results, but does observe that "at some point, we will have to get the kind of leadership that can figure out how to improve our public schools system so that we have the education we want for our children."[12]

Another continuing criticism of standardized testing is that it often favors students from certain socioeconomic and ethnic groups, while disadvantaging others. It is ironic that principals of successful charter schools such as Oakland's Ben Chavis and Jorge Lopez, both administrators in schools with large minority student populations, have no arguments with the testing requirements of *No Child Left Behind*. Chavez asserts that "for the first time, public schools are held accountable for providing minority and poor students with the proficient academic skills needed to compete in our democratic society."[13] Both principals agree, nonetheless, that it is the freedoms extended by their charters that allow them to demand this level of achievement from their students. It is local desire to achieve that makes the difference, not a nationally imposed testing expectation. *No Child Left Behind* simply demands some demonstration of improved performance and does not begin to address the problem of how to get all schools to the point where the majority of students demonstrate proficiency.

The Default Position is Inertia

Because of its representative nature, the power of special-interest groups, and the natural desire of elected officials to accommodate everyone, the federal system moves us toward a position of centrality and equanimity, rather than one of excellence. If we look back at the list of successful federal reforms related to education, they are largely reforms of inclusion—of extending opportunity. As noble and desirable as these objectives are, at the same time they serve to separate and classify by requiring that people identify themselves as belonging to this group or that, to establish how they should be treated. The very word *bureaucracy* draws upon the same root as bureau—a desk with many drawers into which things can be neatly filed and categorized. It speaks to specialization, following fixed rules and policies, and to hierarchy of authority. This is not the stuff of experimentation and innovation. Change is too difficult to monitor and track at the national level, resulting in systems that are plodding and in policies that discourage unique approaches.

Several colleagues recently applied for and received grants from the U.S. Department of State to bring college students from various parts of the world to the United States for eight weeks of intensive English language training and for an immersion experience in American culture. Five of these grants were funded, but the one submitted by these colleagues was remarkably successful by virtually every measure. The students were delighted with the experience, showed significant improvement in English skills, and those overseeing the grant in Washington were openly laudatory. With minor edits on my part to disguise the location of the participating U.S. college, a message to the State Department Coordinator from James Coffman who directs the Malaysian-American Commission on Educational Exchange and assisted with student selection for the language program noted:

> The nine Malaysian participants in the recent UGRAD IELS have submitted their reports and gave a long oral report with photos at the MACEE office on Wednesday. They were absolutely delighted with their program! They had nothing but praise for everyone..., their program counselors and mentors, their teachers, and you as well. For them, as well as for me, the program's quality and breadth of activities went well beyond expectations.
>
> I must admit that during recruitment last fall, I was a little skeptical about what such a short program could accomplish, as well as about sending a group of Malaysians to a frigid part of the U.S in winter. Well, I stand corrected. These students had the time of their lives. What we MACEE staff noticed was not only how much their English had improved, but the greater poise, self-confidence, and maturity they exhibited compared to before their departure. They had really blossomed! And they were all very impressed with the U.S. and American society. Some are now talking excitedly about pursuing their studies in the U.S.
>
> I extend my congratulations and warm thanks to all those who worked hard to make this such a success. Let's do it again next year![14]

The program and its execution were so successful that the proposal written by this particular college was essentially used as the guideline for proposals for the next round of grants. But when those who had initially written the proposal called about reapplying, they were told that they would not be eligible the following year. Grant recipients, no matter how successful their projects had been, could not apply in two successive years.

"How about if we apply through one of our partner institutions who worked with us on making this such a success," they asked?

"That won't work either," they were told by the Washington staff. "There is huge political pressure to pass these opportunities around. Everyone needs to have a chance at them."

"Even if one has shown itself to be particularly successful," my colleagues asked?

"That makes no difference," they were told. "The important thing is opportunity to participate."

In fact, much of the classifying that has become a part of federal requirements is designed to prevent "making distinctions between" in terms of treatment. And, whether we like it or not, to promote excellence is to make distinctions based on ability, effort, approach and choice—not solely opportunity. When a political system by its nature works to minimize the distinctions created by ability, effort, approach, and choice, significant reform aimed at creating performance excellence will not occur through that political system. While we cannot in any way curtail opportunity to participate, if we are to reenergize an education system that has slipped into mediocrity, the solution must develop ways to reward outstanding performance.

True National Standards

There is, however, an important role for our national government in education reform. Continuing in the vein of *No Child Left Behind*, we need national standards of performance that outline essential learning expectations for all students receiving a high school diploma. A study of successful international models that will be discussed at some length in chapter 6 indicates that even though schools in other successful countries are given significant latitude in *how* standards are met, nationally established standards of learning serve the purpose of bringing uniformity to regional curricula, and accomplish the objective noted by Principal Chavez. "Public schools are held accountable for providing minority and poor students with the proficient academic skills needed to compete in our democratic society." Perhaps one of the reasons racism and prejudice maintain their pernicious grip on American society is that we have failed to create a national set of achievement standards that allow all groups to demonstrate equal academic proficiency if provided with the right local support. Absent that national standard, schools are robbed of some of their ability to look outside of their states for models of excellence that can be emulated, since state testing requirements do not provide comparable results. The problem with *No Child Left Behind* is not that it imposes standards, but that they are not actually "standard" or "excellent."

To a large degree, state politics, where the assessment measures are being created, suffer from the same limitations that affect federal government. Since a state government is both democratic and political, it depends on negotiation and compromise to accommodate a host of special interests, and then it depends on the bureaucratic codification of these compromises to govern public systems. Once in statute, state policies become the guiding principles for departments and for public employees whose livelihoods depend on keeping them stable, rigidly enforced, and difficult to change. Despite the volumes of legislation that are created over time, state government remains very much the same. There also seems to be the same seductive appeal about holding public office at the state level, and the desire to be reelected discourages officials with broad constituencies from addressing controversial issues.

As a result, we can know that our Social Security, penal system, and Medicaid system are headed toward insolvency and that our educational system is failing, but few public figures step forward to champion reform, and none are able to gather the needed support to enact significant change at either the national or state level. I am reminded of a conversation with an official in Thailand's Ministry of Education about how much more rapidly the Vietnamese community colleges were progressing in their development than were the Thai colleges. He gave me a wry smile and said, "That's because they aren't a democracy. They can decide something should happen, and make it happen. Democracies are designed to protect the status quo, not bring about change."

State politics does, however, have one characteristic that suggests how educational reform must happen in the United States. Legislatures love to push difficult issues and decisions back to the electorate or to the local level. There is, in fact, an element of genius in our system of government. It allows local government to be about as good as it wishes to be, and local schools to be as strong as they have the will to become. We have been conditioned to think that solutions to major social issues must come from state or federal edict while, at least in the case of education, most of the tools of reform are in our own hands. Given the greater freedom that chartering legislation could afford, an educational revolution could begin in every school district in America, as community after community decides that it will no longer settle for mediocrity, and that its students will be prepared for whatever eventualities the future brings their way. As many countries in Europe and Asia have found, the United States needs to establish rigorous national standards for education achievement, then empower its schools at the local level with the tools necessary to meet those standards.

CHAPTER FOUR

Getting Someone to be Responsible

We recommend that citizens across the Nation hold educators and elected officials responsible for providing the leadership necessary to achieve these reforms . . .

—A Nation at Risk, Recommendation E

When Ying Ying Yu was in grade school, her family immigrated to the United States from China. Once into her teens and attending school in Princeton, New Jersey, she remained deeply influenced by the values and expectations of her early Chinese upbringing. While studying Confucius in her American school, Yu wrote an essay for NPR's *This I Believe* in which she described the personal sense of duty she felt to both her family and the country of her birth—a duty that demanded personal excellence. She began her essay by saying, "I grew up in China, a country where education is the center of every child's life and a grade less than 85 percent is considered a failure." Yu explains that duty and obligation to parents, grandparents, teachers, and country govern her life, and that academic success is a shared honor, not a personal achievement. She concludes her essay: "Only duty will offer me something true, something worthy of my effort and the support of my family and country. Duty can bring me to an achievement that is greater than I am."[1]

When I visit schools or talk to families and students in Asia, I hear the words of Ying Ying Yu repeated over and over as parents, students, and school administrators express this same sense of duty to excel—a duty that is deeply imbedded in the Confucian tradition that permeates much of East and Southeast Asia.

Duty and Responsibility

Vietnam's first university is only a comfortable walk from the tomb of Ho Chi Minh in central Hanoi. It was founded in the eleventh century and still honors the ancient Chinese sage whose teachings formed the basis of its curriculum. Statues of Confucius and his principle disciples fill one of five beautiful halls that adorn the Temple of Literature, and the names of the University's distinguished alumni are etched in stone around the courtyards. Here early scholars were trained in the five Chinese Classics and in the Confucian principle of *li*, seeking to become chun-tzu, the superior, noble, or fully human person. *Li* can be roughly translated as "propriety," "proper comportment," or "correct ritual" and is the principle subject of one of the five Classics, the *Li Chi*.

Of *li,* Confucius said:

> What I have learned is this, that of all the things that people live by, *li* is the greatest. Without *li* we do not know how to conduct a proper worship of the spirits of the universe; or how to establish the proper status of the king and the ministers, the ruler and the ruled, the elders and the juniors; or how to establish the moral relationships between the sexes; between parents and children and between brothers; or how to distinguish the different degrees of relationships in the family. That is why gentlemen hold *li* in such high regard.[2]

Like many Chinese scholars of his day, Confucius was concerned with how to create and maintain harmony in the *Tao*, the way of the universe. His study of the ancient classical writings, particularly the *I Ching*, convinced him that this harmony depended on *li*—on a clear understanding of key relationships among people and of one's duty to conduct one's life in ways that appropriately recognized and honored these relationships. Continuous learning was central to clearly understanding one's duty.

In the *Analects*, the collection of Confucius' teachings, the sage said to a disciple named Yu (a convenient coincidence, considering that he provided the inspiration for Ying Ying Yu's essay!):

> Yu, have you ever been told of the six Degenerations? Tzu-lu replied, No, never. (The Master said) Come, then; I will tell you. Love of goodness without love of learning degenerates into silliness. Love of wisdom without love of learning degenerates into utter lack of principle. Love of keeping promises without love of learning degenerates into villainy. Love of uprightness without love of learning degenerates into harshness. Love of courage without love of learning degenerates into turbulence. Love of courage without love of learning degenerates into recklessness.[3]

To Confucius, learning was at the heart of becoming *chun-tzu*, the superior person, and to the tradition that shaped the educational system Ying Ying Yu refers to in her essay. Confucianism has its own "three R's," Relationships, Respect, and Responsibility, and every person emerging from an education system that is based on Confucian teachings understands these principles. As a result, within these systems, *everyone* accepts responsibility. Schools take great pride in demonstrations of academic excellence. Teachers feel great responsibility for the success of each student. Families pride themselves in how well children perform in school, and students view it as a personal duty to excel. There is, in essence, a complete culture of education.

This commitment to excellence in education takes on particular significance when one considers that there are 1.3 billion Chinese, approximately 330 million of whom are of school-age—more than the entire population of the United States. If we were to divide this group into quintiles, there would be 66 million students in the top 20 percent of this highly competitive student population—more than the total number enrolled in K-12 education in the United States. And *all* feel responsible.

We Are Response-able

Yet it is practically impossible to have a discussion with a group of educators, employers, or parents about academic quality in the United States without getting into the blame game. Employers blame colleges for failing to prepare graduates for the real world of work. Postsecondary institutions blame high schools for producing students who are academically underprepared. Secondary schools blame middle schools, who pass along this fault to the grade schools or to dysfunctional families. And everyone blames society as a whole. No one, it seems, is responsible for why our students do poorly. And this is why reform in education is so difficult and must begin at the community level, with communities who are able to shape the nature of their own educational futures, who choose to be responsible, and are committed to creating a culture of education. If given the freedom and responsibility to make the critical decisions that can result in successful schools, there will be no one else to blame.

I recall listening to a sermon years ago, delivered by someone I have long since forgotten, that contended that responsibility should be spelled *response-ability*, serving as a reminder that, in large part, being responsible is recognizing our own ability to respond. Phillip Howard reminds us that Alexis de Tocqueville and other early observers of American society noted that it is part of the great American tradition to "know how to get together to make things happen"[4]—to be able to respond. Howard contends, however, that ordinary Americans feel that they have lost this power. We are too burdened with legal fear, and too imbued with mistrust. He harkens back to a pre-1960s era when we were much more likely to undertake collective action for the common good, believing that we could take responsible action and benefit from the consequences.

We continue to be bombarded by rhetoric from both the political right and left, reminding us that we must return to or embrace some set of "American" values. For those on the right, these are often defined in terms of religiously prescribed morality, while on the left, values have come to mean embracing difference in any of its manifestations. Howard and de Tocqueville, though, are referring to the values I remember being part of my upbringing in small-town America—values that my international friends still associate with the American promise: independence of thought, ingenuity, integrity, the rewards of hard work, personal achievement, and never settling for less than one's personal best. These are values that are in no way restricted or limited by any particular sense of morality, nor by how we embrace diversity. They are what we would all like to believe are part of "being American." Perhaps these were the values the authors of *A Nation at Risk* had in mind when in 1983 they suggested it was again time to become *response-able*.

Who Else Should We Hold Responsible?

The fifth recommendation in the *Nation at Risk* report specifically instructed local school leaders to step up to the plate and again assume leadership in bringing about the changes needed in American education. Specifically, the report recommended, "Principals and superintendents must play a crucial leadership role in developing school and community support for the reforms we propose, and school boards must provide them with the professional development and other support required to carry out their leadership role effectively."[5] Though very much to the point, this recommendation is essentially backward. School boards hire superintendents and principals, and these administrators respond to the policy directives given by their boards. If students, teachers, administrators, and even families are going to assume responsibility for raising the levels of achievement, our national revolution in public education must begin with school boards. And we are *response-able* when it comes to electing school boards.

In recent years a number of writers have challenged the usefulness of locally elected school boards, arguing that they are ineffective, politically motivated, and often completely dysfunctional. Noted educator Chester Finn commented, "School boards are an aberration, an anachronism, an educational sinkhole. Put this dysfunctional arrangement out of its misery."[6] Several studies have shown direct correlations between the skills of school board members, their propensity to meddle in the day-to-day details of school operations, and performance of children—giving some credence to Finn's concerns.[7] A Virginia superintendent with whom I visited spoke wistfully of the days less than twenty years ago when boards were appointed, remembering them as times when boards were less inclined to meddle in discipline, curriculum issues, and routine administrative matters. By inference, if elected boards have a significant effect on student performance and students are performing badly, boards must be failing. But a number of the nation's largest cities have moved to appointed boards rather than elected boards in an effort to create a governance body populated with talented people who hold the best interests of children in trust, and many of these cities have some of our least successful schools.

I have no argument at all with those who would like to see another form of school governance, and one of the best examples of public school reform in the country at the time of this writing is the Washington DC schools, where a chancellor is being given broad latitude and full support by DC officials who also serve as the school board. But the principles outlined in this book will work whether boards are elected or appointed—if the board chooses to act responsibly. For the reasons I see state governments as unwilling to lead school reform efforts in significant ways, I see states as unlikely to change the governance system for K-12 education in the foreseeable future, and school reform cannot wait. We must begin the process with the governance system we have, strengthened by insistent community support for change.

In thirty of the fifty states, all school boards are elected by local constituents. In every other state, with the exception of Hawaii, all but a few school boards in large urban districts are locally elected. (In Hawaii, there is only one school district for the entire state, and it is governed by the State Board of Education—which is also publicly elected.) In cases in which boards are appointed in large cities, such as Chicago, Los Angeles, Boston, Philadelphia, Cleveland, and New York, mayors and city councils typically appoint these boards. We might look to these city schools to see how successful the appointing process has been and consider the possibility that even there politics gets in the way of bold and effective governance.

In Kansas, only one board is not elected by local residents—the board at the military installation at Leavenworth where the commander makes the appointment. In other words, over 99 percent of the school boards in the United States are publicly elected, meaning you and I decide who is going to be on the board, what they will advocate, and what kind of public support they will receive once in office. In virtually every school district in America, we exercise choice as to how good our school will be. One might argue that legislators are also elected, and I have just made the case that as elected officials their motivation to be reelected and to be publicly recognized prevents them from making difficult and potentially controversial decisions. But at the school district level, few boards are paid—removing some of the incentive to be reelected—and if the community decides that good schools are a priority, board members can do exactly what their constituents wish, and still lead a reform agenda. The same studies that have shown how damaging bad boards can be to school performance indicate that strong boards can contribute in important ways to school improvement.

Establishing Board Values

The Ysleta Independent School District in El Paso, Texas, has 44,000 enrolled students, 91 percent of whom are Hispanic. On the 2008 Texas assessment of student proficiency (TAKS), students at Ysleta ISD outperformed state averages in three of the four testing areas, falling 2 percent behind the state average in Science. (The previous year the school had exceeded the state average in Science by 9 percent.) The average on all tests for the school was 79 percent, compared to a state average of 72 percent, and Ysleta's Hispanic student population matched or outperformed the school average in all subjects.[8] More impressively, between 2007 and 2008, in every testing area but the one noted in Science, the district had improved by between 8 and 11 percent. The district's Web site notes under its History and Mission section that in 1990 its schools were barely meeting state minimum standards, but then the "Ysleta Miracle" occurred. As the district states it, "The major ingredient is no surprise; it is simply treating all children as individuals with the innate ability to learn and providing our teachers and support staff the guidance, support and encouragement to succeed."[9]

A key ingredient to this miracle was a commitment by the Board of Education for creating a culture of education for the Ysleta District. The board's "Core Beliefs" for 2008–2009 state:

1. The needs of children must be placed above the wants of adults.
2. Each child who graduates Ysleta ISD should be fully certified to pursue his/her next career choice whether it be college, technical school, the military, or the work force.
3. All programs should be proven effective, challenging, and research based.
4. Every child and staff member deserves a safe and secure educational environment focused on learning that is free from disruption, bullying, and intimidation.
5. Children deserve to have all employees model appropriate behavior that the student is expected to follow.
6. All students and employees deserve a positive, nurturing environment.
7. Every child is entitled to a teacher that is the best we can hire for that position, so all District positions should be filled solely on merit.
8. Successful schools are the responsibility of the entire community.
9. The District's system of internal administrative and accounting controls should be maintained with transparency for all.
10. The District will provide and promote student opportunities and extracurricular activities that will enhance academic achievement and social development.
11. Children who are healthy and well-nourished learn better.
12. All publics who interact with Ysleta ISD should be treated with a positive attitude, made welcome, provided the information they need, and be treated with respect.
13. Every employee is a valuable team member and plays a vital role in our success.[10]

The evidence is that this board acts according to its stated core beliefs, and that when public school boards commit to doing so, remarkable improvements occur.

Granted, school boards are governed by state policies that dictate what the qualifications of superintendents, principals, and teachers will be, what constitute minimum curricular requirements, and who will be tested in which academic years. The exceptions are charter schools, which are often given much greater latitude, supporting the need to extend these same freedoms to every district. But we are getting ahead of ourselves. First, we need a committed school board with the vision demonstrated by the Leopold and Ysleta Districts, and the will to support radical change. This will require a small group of influential but equally committed community leaders who are willing to plan and direct change on the local level—a Committee on Reforming Education, or CORE Team.

Electing the Right Board

The first time I had to lead the passage of a bond election as a college administrator to fund a college building project, our local Director of Economic Development, who had gained a reputation as one of the most successful community development leaders in the state, taught me an important lesson about elections—and one that is something of a no-brainer once you think about it. In his earlier days, Gib Garrow served as national president of the Jaycees and is an experienced electioneer. As I fretted over the limited advertising budget I had available to try to convince local voters to support the college issue, he said "You know, you don't need a majority of the voters to support your issue. You just need to get a majority of those who vote. Find out what the highest turnout has ever been for a bond issue in this community, add in a fudge factor in case something encourages a record turnout, and you'll know how many votes you need to win. It's a lot smaller number than you think." One of Gib's Fundamental Laws of the Political Universe was that "Many American's don't vote." And if we were to add a corollary to this law, it could be "Even fewer Americans vote in school board elections, unless they occur on a general election day." Add to this the reality that it is often difficult to get people to run for the school board, and you have the perfect formula for getting the *right* people elected.

From Gib I learned a number of other critical lessons about elections and voters. One was that people strongly opposed to an issue are much more likely to vote than those who favor it, and any campaign can generally count on 20 percent of voters to be opposed, no matter what the issue. To get out the positive vote, planners need to create an equally passionate sense of support and a feeling of personal commitment to the cause among a greater percentage of the population. To pick his brain about how a CORE Team might function, I met with Gib to ask how he would identify and stimulate the Team to be passionate about improving community schools.

"Select eight or ten people who are community opinion leaders, then educate them," he said. "Send them in groups of three or four to visit the best examples you can find of the kind of school you want the community to have—like your Delta College Prep or Ysleta in Texas. Let them see that it can be done—and how it is being achieved. They need to become true believers."[11]

This group then needs to draft a plan for what the community's school can become—not simply a generalized "vision" statement, but a specific description of what students should learn, and how the school needs to be structured to achieve those ends. From among its own ranks or from a group of associates, the CORE Team can then find people who will run for the school board committed to this plan of reform. If new to the committee, candidates should also visit examples of excellent schools and become passionate advocates for positive change. This group must become the responsible—and *response-able*—"lead group" for the local reform initiative, working initially in the background to shape the community's school board into committed, courageous, and visionary champions for educational excellence. The process may take some time, since restructuring a board may require several election cycles and the CORE Team's candidates must be politically astute enough to be patient about pushing an agenda that lacks majority support until a sufficient number of reform-minded members are in place to carry the vote. Once that majority is established, it becomes the responsibility of the CORE Team and the seated board to maintain that majority and to support committed leadership in the school while they collectively cultivate a culture of education in the community. As Core Belief Eight of the Ysleta School Board affirms, "Successful schools are the responsibility of the entire community."

Finding the Right Board Members

Despite the naysayers, I prefer to believe that good boards can be elected, and can make judgments that will direct and support a reform agenda. Special-interest groups have demonstrated for years that they can successfully run candidates for school boards and get them elected. Those with a special interest in excellence in education should be able to do the same.

In addition to supporting strong reform candidates, the CORE Team will need to watch for and expose candidates who have conflicting agendas. There will be "single issue" candidates who want to rid the school of a principal who reprimanded their daughter for using obscene language in class. There will be "special agenda" candidates who want to insure that creation science is taught in all biology classes, or that no bond issue is ever put forward by the school district. Others will run as representatives of specific constituencies—the teachers' union or the football booster club.

Some who have been on the board will show themselves to be micromanagers, dropping in on classrooms to evaluate teachers, or wanting to make decisions about textbooks and bus routes. There will also be board candidates who are simply unequal to the assignment—who lack the experience, education, or native ability to deal with the difficult and complex issues involved in bringing about constructive change. In many states, all that is required of a board candidate is that the person be a legal resident of the district and eligible to vote. When election time approaches, the CORE Team needs to remind the public that the role of the board is to develop policies that promote a system that produces graduates who can compete with any in the world, and that these side agendas or other limitations distract from that purpose.

The college board in which I worked for many years, a very progressive and visionary group, adopted the unwritten practice of encouraging members who were not going to run again in the next election cycle to resign their seats four to six months before the end of their term, since the law allowed boards to appoint members to vacant positions until the next election. The board was able to select individuals who were supportive of the college's mission and direction to serve for part of a term prior to running for the seat as an "incumbent." The process worked beautifully, and provided continuity in strong, committed governance over time.

Is this engineering of an election? Only to the degree that all elections in the United States are engineered. Candidates are chosen based on their positions and appeal, and special-interest groups find their own champions and get them on the ballot. Campaigns design elaborate strategies to get the right voters to the polls and promotional materials shape the issues to attract voter interests. What greater cause can we have as voters in a free, democratic society than to identify, elect, and support candidates to our local school boards who will govern with a unified commitment to promote academic achievement and success? Nothing prevents us in our communities from becoming *response-able*, able to decide that we are going to create schools that pass along, from one level to the next, students who have mastered a body of knowledge that prepares them to live fulfilling lives and stand shoulder-to-academic-shoulder with the very capable and highly motivated Ying Ying Yu's of the world.

CHAPTER FIVE

Finding and Supporting Great Leadership

Leadership and learning are indispensable to each other.
—John F. Kennedy

An associate who is vice president and chief legal counsel for a large multinational corporation was asked recently to mediate a dispute that had developed between the board, parents, and superintendent of a charter school near his home in one of our western states. The school had been created in an affluent community by a group of parents who wanted their children to be prepared for admission into the nation's leading universities, with the superintendent hired with explicit instructions to create that learning environment. My lawyer friend had no personal connection with the school and agreed to meet with the parties to listen to their positions, but reserved the right to decline the role of mediator until he knew more about the circumstances. As he spoke with representatives of the three divisions, it became apparent that the new superintendent had done exactly what he was hired to do—design a rigorous curriculum, hire dedicated and capable teachers, and insist on performance from students. Problems developed when some of the students proved unwilling or incapable of meeting the new expectations—at least at a level that gave them the perfect grade point averages expected by their parents. The majority of the board, all elected from the parent group, supported the parent complaint that school leadership was being unduly demanding, and asked the superintendent to lighten up on performance expectations. His response was that he had been hired to prepare students for admission into highly selective universities, requiring not only sterling grade point averages but exceptional test performance. In addition, he assumed parents wanted children to be successful at these universities, once admitted. He had created the school they asked for. The better students were doing very well, and to "lighten up" might produce a class of students with grade 4.0, but not the scholars he was asked to develop. He could not, in good conscience, modify the school's standards.

Before the attorney was able to determine whether he could successfully mediate a solution, the board chose to dismiss the superintendent and began the process of seeking new leadership. The parents and board did not, it appeared, want an exceptional school as much as they wanted the illusion that all of their children were exceptional. Herein lies perhaps the greatest danger school reform faces at any level—a lack of will by parents and by the governing body to find great leadership, then support change through the difficult and often personally challenging process of making the leap to excellence.

Demanding Effective Leadership

Chester Finn and his research colleagues note that, as charter schools are formed, "Zealous parents often have difficulty yielding the reins to professional educators. Nervous, type-A charter boards tend to micromanage...The most common symptom of governance trouble is the abrupt replacement of the principal during the school's first year or two."[1] The tendency to micromanage and to allow disgruntled parents to guide policy is not limited to charter schools, rather, it plagues K-12 education throughout the United States. The St. Louis City School District managed to work its way through eight superintendents in six years between 2003 and 2009, and across the state, Kansas City's urban school district had twenty-five CEOs in thirty-nine years.[2] According to a survey of superintendents in sixty-five of the largest urban districts in the country, the average tenure of these administrators in 2006 was 3.1 years—up from 2.3 seven years before, but hardly sufficient time to initiate broad-based reform![3] In many cases, these leaders were talented, courageous men and women who were unable or unwilling to survive the micromanagement, political maneuvering, or personal agenda-driven actions of board members. But even when a board is properly motivated and behaved, it must have a talented administrative team to initiate and sustain significant reform.

Ron Edmonds, the founder of the Effective Schools movement in the United States in the 1980s, noted the critical role played by leadership in school success by observing, "there may be schools out there that have strong instructional leaders, but are not yet effective; however, we have never yet found an effective school that did not have a strong instructional leader as the principal."[4] Although it is difficult to separate the role of the principal from the many other factors that influence effective school performance, several studies are beginning to shed light on how these chief building administrators make a difference. A study of principals in Florida's Miami-Dade School district, performed by the Institute for Research on Education Policy and Practice at Stanford, found that effective principals are those who manage to attract effective teachers and get rid of ineffective ones. A Texas study conducted at about the same time revealed that the difference a successful principal made on student performance was most pronounced in schools serving students from largely disadvantaged backgrounds. Both studies concluded that good leadership matters, and matters in dramatic ways.[5]

Debate continues to rage in academic circles over whether great leaders have innate talents that enable them to envision, motivate, organize, direct, and inspire more capably than the mediocre or ineffective, or whether these attributes can be learned. Having worked for decades with leaders in education and having studied and taught leadership for a portion of that time, I have reached the conclusion that there are some skills that can be learned or honed, but great leadership requires talents that, if not inborn, are a reflection of a lifetime of learning, experience, observation, and trial. They include abilities that cannot be acquired in the short term from a graduate program in administration, or from a set of certification requirements imposed by state departments of education. Unwavering personal courage, passion for the work based on a set of firmly held values, an ability to relate well to and inspire others, and a talent for analyzing and synthesizing possibilities and selecting those with greatest promise are skills that, if absent in our mature adult personalities, rarely develop.

Leadership Without Certification

Many graduate programs acknowledge the need for these talents but do little to develop them, tacitly acknowledging that they are genetic or developed over a lifetime of experience. As I talk to colleagues in leadership education, most admit that they can tell which students in their programs will be most successful as leaders, and acknowledge that it has little to do with how well each has mastered the academic content of the program. A faculty colleague who teaches courses in K-12 administration noted that he creates group projects in his classes in which he allows the groups to select their own leadership, but confessed that, nine times out of ten, he could choose in advance the leader the group will select. "Some people just exude confidence and a 'can do' spirit" he said. "These are the people the groups choose as their leaders, not necessarily those who are the academic standouts."

Yet most programs in school administration approach these intangibles in only a peripheral way. A review of a number of randomly selected certification programs for superintendents across the country indicated that requirements commonly include a master's degree (often in Education), a qualifying score on a state assessment exam, two or three years in education administration, and continued coursework in K-12 leadership. Required courses typically cover district administration, school law, personnel management, professional development, managing change, collective bargaining, buildings and site management, public school finance, curriculum design, public relations, and supervision of various educational divisions—often accompanied by an internship. All of these address important skill areas for those managing schools, but none makes a student a competent leader.

Thirty-five states have adopted six standards utilized by the Interstate School Leaders' Licensure Consortium (ISLLC) that take a stab at defining some of the affective qualities that support great leadership, including visioning, nurturing a supportive school culture, collaborating effectively, and manifesting personal integrity and ethical behavior. Graduate programs designed to prepare students for licensing exams based on the ISLLC standards utilize case studies and problem solving scenarios to demonstrate these principles and assess students' understanding of them,[6] the same evaluative approach used on most state certification exams. In essence, we are assessing leadership talent by determining if candidates know how to skillfully write about it.

At the invitation of a large national education association, I became part of an exercise several years ago to identify the characteristics needed for successful educational leadership in community colleges. I joined a dozen CEO colleagues in Washington where three other groups had been assembled, consisting of the chairs of the Association's affiliate commissions (people who were also actively involved in leadership positions), faculty of university graduate education leadership programs, and representatives from colleges that had developed "grow-your-own" leadership institutes. As part of the process, each group was asked independently to list and prioritize characteristics believed to be most critical to effective leadership, with the group lists then combined to create a consensus description. As each group reported on its deliberation, I noted that *personal courage* was near the top of three of the lists—but completely absent from the fourth. The list from which courage was missing was the one compiled by university leadership faculty—the group that had been "teaching" leadership for years, but that in many cases had not held top leadership positions. Those who had earned their stripes in the trenches recognized that there was a set of skills that superseded the ability to create and manage a budget, develop a master plan, or effectively delegate. Of course, these management skills were on the final list, but were accompanied by traits such as being entrepreneurial, projecting confidence, communicating effectively, catalyzing involvement by others, exhibiting passion, demonstrating creativity and vision, being courageous, and self-assessing through personal reflection. Even while directing one of these leadership programs, I have become more aware that students can learn *about* these traits, and perhaps enhance them somewhat. But when it comes right down to it, among those who have learned the same content and who exhibit the same understandings, some become very capable leaders and others do not.

This is certainly not to suggest that those who have completed leadership training or certification programs are less capable of being effective or that these training experiences have no value. But many of the distinguishing talents that shape capable leadership can be gained elsewhere, and schools should be free to seek this talent without the limitations imposed by certification. It is difficult, in fact, to judge from a stack of applications who will lead effectively based on degrees, certifications, or listed experience, and institutions are best served by identifying the characteristics they want, then aggressively recruiting or promoting those who exhibit the desired talents. To develop a pool of capable, reform-minded candidates, this search cannot be limited to those who have specific degrees or

who have completed a required state certification program, again neces-
sitating extension of greater freedoms to school districts to hire those with
nontraditional backgrounds.

The remarkable success in improving student achievement demonstrated
by Michelle Rhee as Chancellor of the District of Columbia Public Schools
is an excellent case in point. With a BA in Government from Cornell and a
Master's in Public Policy from Harvard's Kennedy School of Government,
one could hardly argue that Rhee was academically underprepared for
a leadership role in public service. But she was not a "certified" school
administrator in the traditional sense, and her three years of teaching experience had been as an uncertified *Teach for America* recruit in Baltimore.[7]
Prior to accepting the DC position, Rhee managed the nonprofit New
Teacher Project, an organization designed to recruit and train new faculty
for challenging school positions. She initially declined the offer to lead the
DC schools, but accepted when the Mayor agreed to grant her broad latitude and decision-making authority—authority she exercised quickly and
aggressively. Teachers and administrators were replaced, schools closed,
and new security and disciplinary policies enacted, much to the chagrin of
many parents and faculty. At the time of this writing, it is still unclear if she
will survive the relentless pressure from faculty unions who accuse her of
being vindictive and arbitrary in her "weeding out" decisions. According
to the Chancellor, she hires and supports talented teachers, and replaces
those who have not shown acceptable performance with student learning—critically needed changes when one has taken the reins of a district
that had one of the highest per-capita student costs in the country, and was
one of the worst performers. To date, the District's school governance system has been true to its word and has supported her initiatives.[8]

After three years of Rhee's leadership in the Washington DC schools,
elementary student proficiency increased by 11 percent in reading from 38
percent to 49 percent, and from 29 percent to 49 percent in mathematics.
At the secondary level, reading proficiency also increased by 11 percent,
with a math increase of 13 percent. Similar performance increases were
demonstrated by special education populations, and by students who were
learning English as a second language.[9] And this occurred in a public
school district where many of the freedoms of chartering legislation were
extended to management of the district as a whole.

Rhee's successes demonstrate that reform leadership cannot depend
solely on passion, courage, and vision to succeed, but must have the freedom and support to make dramatic changes reasonably quickly. An organization that is in the throes of radical change requires effective fiscal
management, shrewd personnel administration, and a detailed knowledge
of state and federal law and regulations. Rhee brought many of these
characteristics with her. Others, she learned quickly on the job or found in
trusted staff members. Boards that are looking for this kind of leadership
can learn a great deal from those who have defined and modeled leadership competence in other venues.

Theoretical Foundations of Leadership

I came into a college presidency with an academic background in political science, international relations, and media studies, and with no senior leadership experience in the state in which I was employed. I learned immediately that although a nuts and bolts understanding of state policies and financing formulas could be learned, it needed to be present the first day in office, and while I launched myself along a steep learning path, the rest of the management team had to provide some of the knowledge and skills immediately required. Both research and practice have demonstrated time and again that effective leadership in any complex organization is a *team* function.

Mary Parker Follett was one of the early pioneers of what is today considered effective management practice and was once praised by Peter Drucker as having "struck every single chord in what now constitutes the 'management symphony.'"[10] Follett was born in 1868 in Massachusetts and spent her early professional life as a social worker involved in the organization and management of homes for troubled children in Boston. Her keen understanding of human nature and of organizational dynamics led to the publication of a series of treatises on management, most notably *The New State* published in 1920. Two principles espoused by Follett can be of particular value to school boards as they seek capable, multitalented leadership. The first, which she called "The Law of the Situation," stated that each situation contains within it the facts and understandings needed to address it effectively.[11] In simpler terms, the answer to every problem lies within the problem. As individuals, we are often unable to recognize this best solution because our biases, training, experiences, and personal shortcomings "frame" the situation in ways that limit our perspective, requiring application of the second Follett principle—*Power-with.*[12]

Power-with acknowledges that everyone within an organization, assuming their role is important to maintaining a smooth-running operation, possesses a portion of the power of the organization. If the absence of that individual—whether principal, teacher, or custodian—means things don't run as smoothly as they should, that person holds power. Follett argued that, just as with light, if we want the greatest output of energy, we should combine the collective power of those within our institutions whenever time, circumstances, and the nature of the situation allow. By doing so we bring to bear the collective perspectives, talents, and understandings of the group, and thereby are much more likely to see the "best solution" that lies within the situation.

The lesson this provides in selecting great leadership for schools is that a single individual does not need to possess all of the talents needed to enable the organization to achieve its goals. The leadership *team* does. The principal or superintendent—the person with whom the buck stops— must possess that set of intangibles that will sustain his or her resolve to champion results-based reform, but the totality of required skills can exist within the team, directed by a leader who solicits input, recognizes and acknowledges valuable recommendations, and capably executes.

Lee Bolman and Terrence Deal outline four general frames or perspectives through which leaders can usefully view an organization and judge its dynamics and internal and external relationships. Most leaders are inclined to favor one of these four frames or perspectives and shape a leadership style reflective of the way they view the organization. Some choose a *structural* approach to organizational understanding and leadership, focusing on formal relationships and assignments, policy, procedure, clear division of labor, and worker efficiency. Others favor a more *human resources* approach, seeking to maximize performance by harmonizing the "fit" between the goals of employees and those of the organization. These leaders often see organizations as existing to serve human needs, rather than vice versa, and apply themselves to fostering the talents and opportunities of those with whom they work.

Another set of leaders thinks of organizations principally in terms of their "political" involvements and interactions, with politics in this case referring to the process of negotiating, competing for, and creating coalitions to influence the allocation of scarce resources. There is an assumption in this approach that differences will always exist between and among various coalitions, and leadership is largely the utilization of various forms of power to bargain for and negotiate the effective use of resources.

Still other leaders approach organizational life and activity more *symbolically*, believing that an institution's culture, rituals, stories, and ceremonies are the tapestry that binds the organization together and provide meaning to work. When one approaches institutional life from this perspective, what actually happens within the work environment is less important than what it means or symbolizes. In school settings, for example, a requirement that students attend graduation is not important because it guarantees a crowd, but because it signifies to students that this event is the culminating celebration of both their own achievements and those of their classmates, and is worthy of their attendance and support.[13]

The genius of Follett's *power-with* philosophy was that she understood that each organization was multidimensional and must be viewed from a number of perspectives. One of my expressive graduate students, an English Literature undergraduate, described the multiple perspectives or *frames* described by Bolman and Deal as being like literary analysis with "the structural frame as the setting, the human resources frame as the characters, the political frame as the plot, and the symbolic frame as the theme. Each of these factors can be looked at separately but always rely on one another."[14]

Organizational dynamics might also be viewed as looking at an elaborate piece of statuary that is contained within a glass case. By moving around the case, the viewer gains a new perspective from each side, and develops a much more complete sense of the whole from each new reference point. But leaders rarely have the ability to walk freely about the case, viewing its contents from all sides. Their experiences, training, abilities, and biases bring them to one side of the case or another, and though they may be able to lean around in each direction for brief glances, the leaning position is not a comfortable one and can't be sustained for prolonged periods of time. They invariably return to their comfortable side for extended activity. From any side of the case, however, a skilled leader can select others who have approached it from different perspectives, and can ask what they see, developing not a perfect representation of the complete object, but one that is much clearer than any single view allows. This ability to create a composite team view is part of the essence of great leadership.

Identifying Leadership Roles

It would stand to reason, then, that the opinions of parents and board members would factor significantly into the leadership equation. And they do—but only within carefully defined limits. To bring about significant reform, particularly on a broad community level, parents must be educated about the value and necessity of offering world-class education, must have an opportunity to provide input, and must be asked for a written commitment to support the school in its new policies and programs, once established. But this "parent education" must include a clear understanding that the board's role is to create policy, hire capable administrators, and exercise general fiduciary and legal responsibility for the school. Carrying out policy, managing budgets, and doing so legally and ethically are responsibilities of the administration. It is often helpful to have members of the board conduct some of these parent education sessions, forcing them to state publicly that "carrying out policy, managing budget, and doing so legally and ethically *are responsibilities of administration.*"

One of the most useful approaches to defining the roles of public governing boards and administrations is that presented by John Carver in a model he calls Policy Governance. Carver's model stipulates that the board's responsibility is to define the policy parameters within which the school administration must operate—the legal, ethical, professional, and performance expectations and boundaries that will be the basis of evaluation. As long as the administration functions within these prescribed limits and achieves agreed-upon performance objectives, the board has a responsibility to stay out of the way—to keep its nose and fingers out of administrative matters. If the administration does not achieve performance expectations or acts illegally, unprofessionally, or unethically, its members should be evaluated accordingly, possibly resulting in loss of position. But to be evaluated honestly and fairly, the administration must be empowered to be the final word on administrative matters, without board interference or pressure.[15] Policy Governance also emphasizes that the board is a unit and has authority only as a unit. Board members have no unilateral authority. Authority comes only from the collective policy decisions of the board.

The exercise of the Carver Model calls for several additional Fundamental Laws of the Universe—in this case, the school governance and leadership universe.

Law #1: The board, not individual board members, makes policy for the school. When policy has been established, it is the board's policy, and should be supported as such by all board members until it is changed.

Law #2: The person selected or elected to chair the board, not its individual members, speaks for the board. When someone from the media or from the greater community has a policy question for the board, the chair serves as public spokesperson and voices the decisions of the board, not his or her personal views.

Law #3: When someone from the greater community (including parents, teachers, students and the media) has a question concerning policy *administration*, these questions must be directed to the appropriate administrative official, even if the question is initially presented to a board member. It is a difficult but essential requirement that board members be willing to say, "this is an administrative issue and needs to be addressed by our administrative team."

Law #4: To fairly evaluate the performance of an administration, it must be allowed to do its job without interference, subtle pressure, or "hovering" oversight. Regular periodic reporting on performance and goal achievement is a must, but in between, the administration must be left to do its job.

None of these laws is easy to follow. Board members want to let constituents know that they are fulfilling their governance responsibilities, and too often this is interpreted as "taking care of issues" when they arise. But the process of educating the board must include helping each member understand that he or she holds the success of the institution "in trust" and that to best govern, the board must select, support, and fairly evaluate the school's leadership. Each member must be willing to say, "This is what we asked the administration to accomplish, and we stand behind its actions as long as it is pursuing our policy objectives," even when some parents decide they aren't too excited about maintaining high standards after all, and when teachers object to selection, evaluation, and dismissal processes that have been board approved.

Development of policy cannot, of course, depend solely on the knowledge and experience of the elected board. During the transition to reform curricula and procedures, there will be a constant interplay between the board and administration, with policy suggestions originating with each group or coming to the attention of one group or the other from outside sources. Before being enacted, major new policy recommendations should be presented to the faculty and staff for feedback, discussed in parent conferences, and reviewed in open community forums. Community-based reform depends on broad community support, and the first responsibility of successful leadership is to create a new belief in what education can be, and must be, that is shared by the community as a whole. But final policy decisions remain with the board, not with a faculty or community referendum, and execution of those policies rests with well chosen school leadership.

There is a scenario I see repeated over and over in our schools that is undoubtedly one of the principal contributors to the belief by some very astute observers and scholars of K-12 education that elected boards will never lead us to significant reform. A district will find a capable leader; a person who has shown great success in previous positions in moving change forward. The new leader is given a mandate to do the same in the new position, and assured that the board is firmly behind her in getting done what needs to be done. But as soon as a school needs to be closed, a group of teachers needs to be replaced, a sports program needs to be dropped to provide resources for a critical academic program, or a disciplinary issue comes to a head, board members begin to fold under the pressure of public criticism or personal desire. So, to conclude this chapter about finding and supporting great leadership, I pose these questions to those who now serve on school boards or are considering running.

Are you a leader in that you are engaged in visioning, nurturing a supportive school culture, collaborating effectively, and manifesting personal integrity and ethical behavior?

Do you possess unwavering personal courage, passion for the work based on a set of firmly held values, an ability to relate well to and inspire others, and a talent for analyzing and synthesizing possibilities and selecting those with greatest promise for improving student achievement?

Can you look a friend and constituent in the eye and say, "I realize that this is not what you would like to see happen, but it is the collective best judgment of our board that we need to move the school in this direction?"

If unable truthfully to answer "yes," how can you expect an administrator from whom you require no less to succeed, when his success depends on your unwavering support? Critics of the elected board system may doubt that we can find through the electoral process five or nine brave and committed souls willing to exhibit this strength as community volunteers. But we must, or the professional leadership we select will fail.

I believe that we are a country rich with these people. They come out of the woodwork when an old growth forest is threatened, or when a community attempts to enact an ordinance that threatens a tenet of their faith. When we decide that our children are as important as our trees, and that their education is critical to maintaining our values and beliefs, we will have on school boards the leadership we need to become an educated nation.

CHAPTER SIX

Developing a World Class Curriculum

> Our nation faces a deeply troubling future unless we transform the lost opportunities of the senior year into an integral part of students' preparation for life, citizenship, work and further education.
> —Lesser and Greater Expectations Report, AAC&U

Just before the beginning of my freshman high school year, my father moved our family of eight to Teheran, Iran, where he had accepted a position with a State Department program designed to provide English instruction and library preservation services to Iranian universities. As a result, my ninth-grade year was spent in an American Dependants' school in the Persian capital, and my tenth, in an international high school run by the Presbyterian Church, where the student body came from every part of the world. In combination, the two schools provided some of the most exciting educational opportunities of my young life. Bess Byrom, my ninth-grade English teacher, introduced us to Shakespeare by coaching our class through a production of the "play within a play" in *A Midsummer Night's Dream*. I still can quote Pyramus' lines by heart, and credit her with a deep and lasting appreciation for English literature. Mrs. Young, who taught sophomore English, was married to an archeologist working in Persepolis. During our reading and discussion of the Epic of Gilgamesh, he visited our class to describe how myths and stories moved across the Middle East as the Persian armies of Cyrus and Darius swept from Babylon to the Mediterranean coast. There was no argument in these grades about whether the focus in these English courses should be on writing or literature. We read and wrote continuously, with each writing assignment graded as if it were an entry in a literary contest. Mrs. Young had an adventuresome spirit and a spine of steel, and I wasn't at all surprised to learn years later in 1985 that she had been a passenger negotiator with Palestinian hijackers aboard the Greek cruise ship, the Achille Lauro. In Mrs. Young's and Bess Byrom's classes, school was fun, invigorating, and challenging. (My older sister remembers these years as grueling, with as much homework as she ever experienced while later attending a prestigious private university, but she is a perfectionist, and I was a fifteen-year-old realist.)

I returned to the United States for a full but uninspiring junior year, only to find that all that remained to meet the graduation requirements of my American high school senior year was a physical education course and an American History class I had missed while abroad. A second year of Chemistry and additional math and English filled a senior schedule, but I learned that even during the 1950s and early 1960s the era my generation hearkens back to as the halcyon years of American education, high school graduation requirements did not demand four full years of substantive coursework. We should not be surprised that, in the decades since then, countries with a strong, rigorous elementary and secondary curriculum have passed us by.

The Wasted Senior Year

Little has improved since the middle of the last century in our efforts to make better use of the senior year of high school. A report by the Association of American Colleges and Universities (AAC&U) noted that, while graduation requirements vary by school district, they "may be set so low that seniors need to enroll in only two or three courses to earn a diploma. The schools themselves are sending a message that the senior year is a time of low academic expectation."[1]

In a longitudinal study of out-of-school factors contributing to low school achievement by adolescents, psychologist Laurence Steinberg found in nine schools in California and Wisconsin that high school students spent 15 hours on extracurricular activities per week, 20 to 25 hours socializing, 15 hours watching TV, and 15 to 20 hours working: making a total of 65 to 75 hours in a 168-hour week.[2] This study is now a little dated, and socializing has been replaced by social networking, and some TV watching replaced by texting and gaming, but if a student were to average only seven hours of sleep and seven hours in school per class day, little time is left for any other activity—including studying. No student with this schedule could manage the homework expectations my sister so strenuously objected to at Teheran's Community High School. This reality led the AAC&U Report to conclude that "schools and parents do not demand that significant time be spent on homework or school projects."[3] Steinberg's observations were that "American high school students do not take school, or their studies, seriously," and, more critically, "American parents are just as disengaged from school as their children are."[4] Although Steinberg's report was published over a decade ago, there is little to indicate that student or parental commitment to school has changed.

International Comparisons

International comparisons of student performance, particularly in areas of math and science, demonstrate the consequences of the "softening" of the U.S. high school curriculum. On the 2007 Trends in International Mathematics and Science (TIMSS) assessments, although American fourth- and eighth graders scored above the forty-eight-country average, fourth graders trailed ten countries in Europe and Asia in mathematics, and eighth graders trailed eight. In Science, U.S. fourth graders trailed seven countries, largely in Asia, and eighth graders scored below the performance of ten other countries. The People's Republic of China elected not to be part of the TIMSS study, so we have no comparison with the performance of Mainland Chinese students.[5]

By the time students reach high school, comparative performance further declines. Data from the Program for International Student Assessment (PISA) study administered by the Organization for Economic Cooperation and Development (OECD) to fifteen-year-old students assesses literacy in both mathematics and science. Students in the United States scored below the international average in both areas, and, in mathematics, twenty-three of the thirty OECD countries and eight non–OECD member states outperformed the United States. In science literacy, sixteen OECD countries scored higher than our students, as did fifteen-year-olds from six non–OECD countries.[6] Finnish students, the top-scoring group, were almost two grade levels ahead of the American students, who ranked closer to Mexico, the lowest scoring OECD country.[7] Again, although Chinese Taipei, Hong Kong, and Macao were included among the non–OECD countries, Mainland China was not.

Comprehensive, Content-Rich Curriculum

An examination of curricular requirements in some of these countries provides much of the explanation for these differences. In 2009, the organization Common Core launched a study of the nine nations (or territories, in the case of Hong Kong) that consistently outperform the United States on the PISA assessment: Finland, Hong Kong, South Korea, Canada, Japan, New Zealand, Australia, the Netherlands, and Switzerland. While these countries differ in significant ways in the approaches they take to achieve high academic performance, they share a number of similarities that suggest what needs to happen in this country if we are to equal their performance. As a beginning point, all have a longer standard school year than do schools in this country, with all but Canada ranging between 190 and 243 days. (Canada varies from 180 to 200, depending on the province.) The typical school year in the United States is 180 days, differing a day or two by state and region of the country. Kansas leads with 186 and North Dakota trails with 173.[8] In essence, students in nations that outperform us are spending ten to sixty more days in school annually than do students in our country. Through twelve grades of schooling, a Japanese student spends the equivalent of *four* more academic years in school than does a student in the United States!

More importantly, the Common Core study found that these top-performing countries provide students with "a comprehensive, content-rich education in the liberal arts and sciences."[9] The report notes that, at the time of the study, while some nations had national curricula, none of the state-based countries being reviewed did. Some countries required national testing while others did not. What they all shared was the sense of responsibility mentioned earlier for providing students with a content-rich, broad-based, internationally competitive education. To avoid suggesting that there is a "model" curriculum that should be emulated by all, the report chose not to outline in detail the year-by-year curriculum of any particular nation, but provided excerpts from each, illustrating its rigor and "content-rich" design.

In Finland, which has consistently ranked first or second in math, science, and reading, students in seventh through ninth grade physics will:

- Learn to work and investigate natural phenomena safely, together with others.
- Learn scientific skills, such as the formulation of questions and the perception of problems.
- Learn to make, compare, and classify observations, measurements, and conclusions; to present and test a hypothesis; and to process, present, and interpret results, at the same time putting information and communication technology to good use.
- Learn to plan and carry out a scientific investigation in which variables affecting natural phenomena are held constant and varied and correlations among the variables are found out.
- Learn to formulate simple models, to use them in explaining phenomena, to make generalizations, and to evaluate the reliability of the research process and results.
- Learn to use appropriate concepts, quantities, and units in describing physical phenomena and technological questions.[10]

Looking to Asia, fourth through sixth graders in South Korea have a set of learning objectives in the visual arts that include:

- Developing creativity and imagination: Through active participation in art appreciation, criticism, and making, students will develop new and different ways to enhance their power of imagination, creative thinking, and presentation skills.
- Developing skills and processes: Students will learn to use visual language, different visual art forms, and a variety of materials and techniques for visual arts making.
- Cultivating critical responses: As students learn to understand works of visual arts, they acquire the abilities to give critical, informed, and intelligent responses based on well-explored background of information about the artwork, the artist, and just as importantly, with

reference to their own experience, training, culture, and personal judgment.

- Understanding arts in context: Students will learn to understand the meaning and value of works of visual arts in their own and other contexts including the art historical, personal, social, cultural, ideological, and political.[11]

Closer to home, seniors in Canada are asked on a high school exit exam in British Columbia to identify on a map of the Near East the countries that were under British mandate following World War I, and to identify the Anglo-Saxon value represented by Herot in the epic tale of *Beowulf*. This comprehensive exam assumes familiarity with some thirty-six literary works, ranging from John Donne to Dylan Thomas. Eighth graders in Ontario are expected to demonstrate music skills that allow them to read music appropriate to their age, identify and play the major scales they find in the music they play and sing, and describe some historical context of the music they normally consume.[12]

Martin West, in an essay written as part of the Common Core report, notes that an examination of the curriculum from these leading countries reveals that "a rigorous curriculum rich in a wide range of subject content matters greatly for students' academic success, over and above the policies and systems of governance that shape how that content is delivered."[13] He speculates that one of the shortcomings of our assessment programs in the United States is not that we are testing children, but that we are evaluating their learning too narrowly and are encouraging states and schools to do so by imposing benefits and sanctions based on narrowly focused criteria. *No Child Left Behind*, for example, requires states to test in science, but does not ask that these science scores be used in determining Adequate Yearly Progress. There is no requirement to assess in history, or in a number of other core subjects at all. As a result, schools in the United States narrow their curricular focus either through what is offered, or through what is emphasized and evaluated. The countries that outperform us do test, but they both educate and assess broadly.

The Common Core report provides clear evidence that top-performing countries utilize curricula that are, by design, highly content-specific in a wide range of subject areas, starting at the earliest grades. Students are expected to master a specific body of knowledge—not just be able to demonstrate skills in synthesis, analysis, and application. Curricular emphasis in the United States, on the other hand, has evolved toward basic "skill building" in reading, computation, and written communication, with decreasing emphasis on the accumulation of a broad, rich, but specific, body of knowledge. Yet performance by students in these leading countries indicates that students who are well versed in a knowledge-specific liberal arts curriculum do better on assessments of math and sciences than do our "skills" prepared students. Perhaps this is a reflection that *content* provides or enhances an understanding of *context*—that students with broad mastery of the events, ideas, and artistic expressions that have shaped our past are better able to see meaning and value in academic skills acquired in the present. There is also an implication in this observation that teachers must be content specialists as well as education generalists, able to provide rich content in such a way that context becomes evident.

National Standards, Local Decisions

Rather than rising to the occasion, recent research indicates that states may actually be lowering their standards for proficiency. Every two years the National Center for Education Statistics (NCES) applies a formula that compares requirements for proficient performance across states. In addition to finding that the difference in achievement requirements between the states with the highest expectations and those with the lowest was as great as the difference between *basic* and *proficient* levels on the National Assessment of Educational Progress (NAEP), the NCES study found that expectations were actually declining. Results compared available data for 2005 and 2007 and found:

> In grade 4 mathematics, 14 of the 35 states with available data in both years indicated substantive changes in their assessments. Of those, 11 showed significant differences between the 2005 and 2007 estimates of the NAEP scale equivalent of their state standards: 6 states showed a decrease and 5 showed an increase.
>
> In grade 8 mathematics, 18 of the 39 states with available data in both years indicated substantive changes in their assessments. Of those, 12 showed significant differences between the 2005 and 2007 estimates of the NAEP scale equivalent of their state standards: 9 showed a decrease and 3 showed an increase.[14]

In mid-July of 2009, the Science, Technology, America and the Global Economy program (STAGE) of the Woodrow Wilson Institute sponsored a roundtable discussion in Washington DC, at which Andreas Schleicher, head of the Indicators and Analysis Division of the OECD, offered his perspective on why U.S. schools trail many of their OECD neighbors. Paraphrased here in a report on the meeting, he indicated that:

> The countries with the best scores on the PISA all have clearly defined and challenging universal standards, along with individual school autonomy. This approach is emulated in most European systems, which have centralized standards or "goals" for how students should perform in each grade level, but give schools discretion with their curriculum, budget, organization, hiring, and teaching decisions. The United States, on the other hand, lacks universal standards and also has a much lower degree of local control than other

countries, leading to both widespread variability in student learning outcomes and lower performance.[15]

William Schmidt of Michigan State University, who at one time oversaw U.S. participation in the TIMSS assessment, agrees with Schleicher and describes the U.S. curriculum as "lacking focus, rigor and coherence."[16]

Schmidt and his colleagues at Michigan State note that, when the 1997 TIMSS results were reviewed, Germany also discovered that it had fallen behind many of the rest of the developed world in educational achievement, and ranked at the same level as the United States. Its response was to evaluate why other nations had done better, and Germany's conclusion was that a set of national performance standards appeared to be a common denominator for most high-performing systems. Over the next decade, despite strong objection from proponents of its "states-based" education system, Germany developed a set of comprehensive national standards, while maintaining local control of how those standards could be met.

Responding to the same discouraging testing results, the United States government under the Clinton Administration proposed a voluntary national test but was unable to overcome our historic commitment to local autonomy, and Congress killed the initiative. The Michigan State study indicates, however, that if we are to compete internationally in the education arena—and by implication, in the economic arena—we may need to find an acceptable balance between developing national standards of performance and retaining local ability to meet those standards.

Learning from the Leaders

Schmidt and his team identified six key lessons drawn from an examination of ten countries that have made significant progress in educational achievement while we have languished. These include:

Lesson 1: It is not true that national standards portend loss of local control.[17]

The Michigan State study illustrated that there need not be a relationship between establishment of national performance standards and loss of local control by schools. In the German experience, local control was actually broadened as the federal government negotiated a compromise that allowed for development of the standards. A similar undertaking in the United States could reasonably create a set of standards by which all schools can be measured, while extending to those schools the opportunities now available only to charter schools to meet those standards.

Lesson 2: Create an independent, quasi-governmental institution to oversee the development of national standards and assessments and produce reports to the nation.

All of the countries involved in the Schmidt study found that it was not possible to develop "focused, coherent and rigorous standards for *all* children"[18] without some form of national standards institute. The study's authors recommend an institute after the German model where a nongovernmental organization (not part of the U.S. Department of Education, in our case) has quasi-independent status similar to that of the National Academy of Sciences and is essentially created by the states to oversee development and assessment of national standards. The report recommends that the Institute could include an apolitical board of academics, educators, public officials, and representatives of the public. Appointments could be made through the states, led by the governors. Specific qualifications would vary for different categories of membership but should generally include subject-matter expertise and/or relevant experience in the field.[19]

In 1996 a collaboration of governors and business leaders formed Achieve, a not-for-profit initiative designed to "raise academic standards and graduation requirements, improve assessment and strengthen accountability." The American Diploma Project, launched by Achieve in 2005 and now including 80 percent of the states, and the National Governors Association's Common Core State Standards Initiative suggest how a model might be developed to create nationally applied standards. At the time of this writing, the Common Core State Standards Initiative had been endorsed by every state but Texas and Alaska.[20] It is worth noting, however, that the NCES study of changes in standards between 2005 and 2007 found that the majority of states in which there were reductions in requirements in what constituted "proficient" performance on eighth grade math exams were America Diploma Project states.[21] If these initiatives or something like them are to take the lead in national standards development, the agreed upon measures will need to have both "teeth" and firm commitments from every state to use them as minimums, while working to rise above them. Quoting the American Productivity and Quality Center, the report by the National Governors Association on benchmarking new national curricula standards acknowledges that: "Benchmarking is the practice of being humble enough to admit that someone else has a better process and wise enough to learn how to match or even surpass them."[22]

Lesson 3: Position the federal government to encourage and provide resources for the standards-setting process.

Under the German model, a nongovernmental institute known as the KMK established the national standards, with the initial development work funded by both the states and the federal government. Participation by states in the national standards program was initially voluntary, but all have now elected to participate, in part because the federal government's financial incentive powers were directed toward encouraging participation, rather than setting and enforcing the standards.

Lesson 4: Develop coherent, focused, rigorous standards, beginning with English, math, and science.

As noted in the discussion above, each study of the TIMSS countries has indicated that those demonstrating highest achievements have focused, rigorous, and coherent curriculum standards, with "focus" referring to concentration of instruction at each grade level around a limited number of key topics, with emphasis on mastery. All of the nine countries that studied with national standards have achieved mastery in math and language arts, and eight have them in science. Development of these standards should be followed by others in history, economics, social studies, the arts, and foreign language.

Lesson 5: Administer national assessments (including open-ended questions) at grades 4, 8, and 12 every two years.

Since the National Assessment of Educational Progress is already administered to these grades every two years, a national standards assessment could reasonably utilize or take the place of these tests and be administered following the same schedule. The international model is to include a variety of question types on these assessments, including open-ended questions that require grading. The costs involved in hiring graders suggest another area where the federal government can play an important role.

Lesson 6: Hold students, teachers, and schools accountable for performance.

Accountability in the countries studied applies to students, classrooms, schools, regions, and the nation. In some countries, students do not progress between grades or graduate from secondary school unless they demonstrate prescribed levels of achievement. Yet graduation rates in many of these countries exceed our own. Teachers and schools are evaluated based on performance, often influencing resource allocation. In Singapore, schools are ranked by results and the results are published. The Michigan State study recommends that, at a minimum, once a national assessment model has been shown to be reliable, twelfth grade results should be used as an indicator of readiness to enter the workforce or move on to postsecondary education.

Critics have argued that many of these international comparisons are invalid, since our egalitarian spirit in the United States extends educational opportunities to all citizens, while many countries limit access exclusively to the elite. Perhaps this criticism had merit in the middle of the last century, when, as a country, we led the world in the percentage of its population with a high school diploma. But we now rank near the middle of the OECD countries in this statistic (12th), and the "opportunity" argument no longer holds water. It is true, as other critics maintain, that we are the most heterogeneous of the OECD nations, with a much more diverse population, and therefore face greater educational challenges. And there is some basis for the argument that diversity affects performance. If we compare testing results on the PISA assessments for Minnesota, for example, with those of OECD countries, the state stacks up very well! Diversity complicates the creation of a common core of knowledge when students from widely varying cultural backgrounds come together in the same school setting.

But essentially this is beside the point. We are a nation at risk, and it is as a diverse nation that we must respond and become competitive. We often excuse the performance on standardized measures of academic achievement by certain groups in our society by arguing that our testing instruments contain bias *because* students lack a common knowledge base—providing even greater support for a rich, comprehensive, content-based curriculum beginning at the earliest grades, and required of all. From a purely pragmatic standpoint, in a world in which the graduates of our schools will hold a series of jobs in multiple careers, working across international and cultural lines, to excuse the ability to perform because an individual lacks the content knowledge and understanding of the social, political, and cultural context required by the job will simply not be acceptable. Our cultural complexity should, in fact, provide an advantage as we design our own content-rich curriculum that acquaints students with the history, literature, music, belief systems, and values of our wonderfully complex society.

Strengthening Core Content

As districts interested in reform seek curricular models that can move them toward parity with these international leaders, they need not look too exclusively abroad. A number of leading curriculum specialists in the United States have designed models that reflect a content-rich approach to learning. The curriculum recommended by the LEAP study and described briefly in chapter 2 made its first recommendation that students must be able to "demonstrate knowledge of human cultures and the physical and natural world through study in the sciences and mathematics, social sciences, humanities, histories, languages, and the arts." This should be accomplished by engaging students "with the big questions, both contemporary and enduring."[23]

Decker Walker, a leading specialist on curriculum reform, noted that at one point the Education Commission of the States listed twenty-six curriculum school reform models on its Web site, along with another eighteen skill- and content-based approaches to curriculum development.[24] Of these, the one that has received greatest attention in recent decades for its focused attention on the need to develop within students a specific body of knowledge is the Core Knowledge curriculum of E.D. Hirsch.

Hirsch, whose book *Cultural Literacy* emphasizes the critical importance of content-specific learning and fostered the Cultural Literacy and Core Knowledge movements in elementary curriculum, notes:

> knowledge of content and of the vocabulary acquired through learning about content are fundamental to successful reading comprehension; without broad knowledge, children's reading comprehension will not improve and their scores on reading comprehension tests will not budge upwards either. Yet, content is not adequately addressed in American schools, especially in the early grades.[25]

A number of successful charter schools, such as Parker Core Knowledge Charter School in Parker, Colorado, utilize Hirsh's Core Knowledge model. Of its curriculum, Parker's Web site declares:

> From the beginning (Kindergarten), we consider it our job to train the disciplines associated with strong academic success. All grades have nightly homework, and students learn to work, think and grow within a highly structured and challenging classroom environment. Core subjects—math, reading, writing *and Spanish*—are strongly emphasized, with physical education, the Arts, sciences, technology and humanities rounding out the students' daily schedule.[26]

Teachers are granted broad latitude in how they teach, but religiously follow the grade and subject-specific Core Knowledge sequence that outlines in detail what is to be taught in grades K through 8 in Language Arts, American and World History, Geography, Visual Arts, Music, Math, and Science. In first grade World History, students learn about ancient Egypt, the African Continent, the Sahara Desert and the importance of Nile River floods to farming, about the Pharaohs Tutankhamen and Hatshepsut, pyramids, and hieroglyphics. By the eighth grade they are discussing the Vietnam War, the rise of social activism, and oil politics in the Middle East.[27]

The Core Knowledge curriculum extends through grade twelve and is only an example of content-specific instruction. As the Common Core study emphasizes, it is a broad curriculum rich in subject content that matters, not a specific delivery method. Those introducing new curricula must also remember that it has been our emphasis on creativity, analytical thinking, and freedom of expression that has made our postsecondary system of education in the United States the international hallmark. Any content-rich curriculum designed to prepare students to enter this system must require students to apply this knowledge in creative and integrating ways.

During a flight to Beijing, I was seated beside a Chinese woman who was returning from a postdoctoral year at the University of Michigan. Her PhD was from Beijing Normal University, considered by many to be the most prestigious in mainland China. When I asked why she had decided to come to the United States for additional study, she smiled shyly and said, "I was educated in China in the 70s and 80s when we sat quietly in our classes and never questioned our professors. My head is full of facts, but I came to the U.S. to learn how to think." We must insure that our own children do not lose the ability to think, while providing an academic foundation that gives them something worthwhile to think about.

Implementing Curricular Change

In the absence of a demanding senior year curriculum, a number of educators have suggested alternative plans for use in the final year of high school, ranging from sending students on to work and college early, to dividing the day between a few remaining required classes and community service projects. There is a strange irony in expressing concern, on one hand, that students are not learning what they need to know in high school, while, on the other, exploring ways to shorten the schooling experience. In an era that is now universally referred to as the "age of information," we should be exploring ways to expand learning opportunities, not shorten them. I fully support service learning, but it should be part of a rigorous curriculum in English, mathematics, the arts, sciences, and social sciences that fills each of the high school years, sending students on to college with knowledge and skills that insure that we not only have the best system of higher education in the world, but also that the best students are entering it.

Curriculum revision is challenging to a school district under the best of circumstances, possibly contributing to Bill Gates' observation that his foundation has had less success trying to change an existing school than create a new one.[28] A number of those who study the charter movement have observed that those starting from scratch have much greater likelihood of success than do those that attempt to transform existing schools. As Izumi and Yan note, "The powerful special-interest players in the public schools system are hugely resistant to real change."[29] Occasionally a district will have the opportunity to open a new elementary school and "pilot" a new set of standards, but unless one is creating a school with new leadership, a new teaching staff, and new procedures, districts will face the challenges of rethinking and restructuring how and what students learn within the constraints of the existing system.

There may be wisdom in selecting one to three grade levels where, following a period of intense teacher education, new curricula can be introduced. Students in these grades can then be tracked as a cohort through progressive curriculum change as they move from grade to grade. If, for example, grades one, five, and eight are selected, over a four-year period a complete program transition can occur in all grades. But I am violating my own advice that the nuts and bolts of school reform activity must be left in the hands of capable administrative leadership. The critical issue for the community is that it must insist upon and support a change agenda, resulting in implementation of an internationally competitive course of study.

Here again, the community CORE Team can play a vital role, as it works in an advisory capacity with the school board. Through a subgroup of distinguished business, public sector, and education leaders, a "curriculum advisory committee" should research, evaluate, and recommend either a curriculum plan or curricular characteristics to guide school policy, focusing specifically on content-based learning and a secondary school program that takes full advantage of all four high school years. Following the suggestions of the Michigan State study for who should be part of a group to shape national standards, this curriculum committee should include not only subject-matter scholars but "others with deep subject-matter knowledge who are engaged in related professional, business and vocational fields." These committees could endear themselves to nearby Colleges of Education by funding research assistantships for capable graduate students who can help with literature reviews, evaluation of curriculum options, and dissemination of information to the community.

Decker Walker notes that "much of the real action in curriculum change occurs outside of the official framework of federal, state, and local policy making.... It normally comes from advocates organized to develop, propound, and secure acceptance of the change."[30] If we are to see significant reform in America's schools, in each community in the United States a CORE Team must work with its reform-minded school board to enact policy that calls for a complete, content-based curriculum. The community must then support the school's leadership as it puts in place and trains a cadre of teachers to present the new content with passion and imagination.

CHAPTER SEVEN

Math as the Language of Science and Technology

"I have heard many People say, 'Give me the Ideas. It is no matter what Words you put them into.'"

—William Blake

During a visit to a model elementary school in a rural community in central Thailand, I asked the principal how many languages her students were required to learn.

"Three," she said without hesitation. "They must learn Thai, because it is their native language. They learn English because it has become the language of international commerce. And they learn mathematics, because it is the language of science and technology."

Her observation was both simple and profound. Just as our letter-based language serves as the means by which we symbolically present and make sense of our social, cultural, and relational world, our number-based language enables us to understand symbolically the scientific and technical world. Yet somehow this lesson was lost on those who struggled to teach me algebra in junior high, and I came away from the experience thinking of mathematics as no more than solving numerical puzzles to see if I could find the right answers. Judging from the aversion many Americans have to mathematics and to the disciplines that rely on it as their "language," many of us had the same experience.

The Need for "Powerful" Math Education

Despite our national aversion to math-related subjects, strong fundamental learning in mathematics and the physical and biological sciences is central to our ability to compete in a world that is increasingly driven by technology. An article in Forbes in 2008 listing the ten most lucrative college majors for those receiving a bachelors degree indicated that the top eight were all mathematics based: four fields in Engineering (computer, electrical, mechanical, and civil), Economics, Computer Science, Finance, and Mathematics.[1] During this same year, Business Administration continued as the most popular major, followed by Psychology, which had jumped by 22 percent from the previous year to gain the second spot. Job opportunities in psychology? There certainly are some, but for those looking for a comfortable income, the Forbes study indicated that psychology graduates ranked second from the bottom, just above criminal justice majors, in terms of expected salary. The key to a high-paying job immediately after college is a solid background in mathematics, yet only 4 percent of college graduates in the United States major in Engineering, compared to 20 percent in Asia and 13 percent in Europe.[2]

In 2007 the Nation's Report Card on Mathematics indicated that math performance for fourth graders in the United States was gradually improving and had risen by 27 points on a 500 point scale between 1990 and 2007 (213 to 240). Nonetheless, a cut score of 249 was viewed as indicating "proficient" on the math assessment, leaving our national average nine points below the proficient level. Eighth grade scores also increased between 1990 and 2007 by 18 points, rising from 263 to 281. Increases occurred for both genders and for all income levels, but 39 percent still tested below the eighth grade proficient level of 299.[3]

Trends in International Mathematics and Science Studies (TIMSS) remind us every few years that students leaving our secondary schools are lagging behind much of the developed world in academic proficiency in these critical disciplines. American fourth graders trailed eight other countries and were tied with four (all in Asia and Europe) in the 2007 TIMSS study of mathematics, while eighth graders were outperformed by five countries and tied with five.[4] In the sciences, our fourth graders ranked behind four countries and were tied with six (all in Asia) and eighth graders fell behind nine countries and were tied with three. By age fifteen, students in twenty-three countries of the thirty OEDC nations were placed higher than U.S. students in mathematics in 2006, and sixteen countries outperformed us in science. Lower achievement in high school leads to fewer students entering degree programs at the college level, and prompted Andreas Schleicher to observe at a meeting of international educators hosted by the National Science Foundation in Washington:

> over the past ten years many countries, including Australia, Sweden and Norway, have caught up and surpassed us in achieving much higher college graduation rates, especially in the areas of science, mathematics, engineering, and technology (STEM). The rapid improvement in access to education and college graduation rates in other nations has significantly contributed to the expanding global talent pool.[5]

Why Do Others Perform Better?

Why are other countries consistently outperforming us in math and science? William Schmidt credits it in part to a difference in emphasis in curriculum. He notes that most countries cover only three or four major math concepts in a given grade, focusing on student mastery of these select concepts before moving on to the next level. The United States, on the other hand, may cover as many as twenty. This approach, which he refers to as being "a mile wide and an inch deep,"[6] provides our students with only a superficial understanding of many concepts, while students in Europe and Asia are becoming thoroughly grounded in fundamental principles that build on each other from year to year. By the eighth grade, students abroad are concentrating on algebra and geometry, while many American eighth graders are still focusing on general arithmetic and fractions.

Schmidt also expresses concern that the curriculum in schools in the United States lacks coherence, with concepts failing to build upon each other from one grade to the next, and no commonality from school to school or from state to state. Limited effort is made to establish this continuity as students progress through the grade levels, and many complete their math education at the end of their tenth grade year.[7] Some concepts important to further math study are never covered at all.

Much of the discussion in the international literature about mathematics education now focuses on helping students grapple with "powerful mathematical ideas." Through the past century in the United States, emphasis in the elementary grades has been on teaching children to perform computational procedures, culminating in high school with manipulation of algebraic symbols. Little emphasis has been placed on helping students understand what mathematics is all about, beyond seeing it as a series of number puzzles. The "new math" of the 1970s and 1980s made an attempt to integrate mathematical ideas into the curriculum, but failed in that neither those teaching these ideas, nor parents trying to help their children learn them, understood the way new concepts were being presented. As one article discussing the failure of new math described it, "The sources for most of the theoretical and philosophical arguments that generated the new mathematical content were mathematicians," and its failure resulted from it becoming no more than "formalistic game-like plays in and with structures defined in terms of sets and logic; often devoid of sense-making relations to matters outside the structures themselves."[8]

Both this observation and the challenge a layman might have in deciphering the quote above illustrate much of the difficulty in helping our students grasp "powerful mathematical ideas." Many students (and their teachers) who say "I'm not good at mathematics" are simply saying "I don't understand mathematics in the way it is being presented to me." If we are to teach students the language of mathematics, we need to find those who can express it in terms others—especially those who are teaching elementary students—can understand. Some might argue that if we are striving to teach children the "language of mathematics," they need to be familiar with the way it is spoken by mathematicians. I would argue, on the other hand, that this is akin to saying that if students are to learn

Chinese, they must master a specific dialect. Certainly students must gain a mastery of the symbols, concepts, and key terminologies that form the essential structure of the language, but it is the responsibility of those who are specialists to be able to define and describe the critical concepts of the language's "culture" in terms that can be understood by someone with reasonable fluency with these symbols.

Teaching Mathematics as Language

Barbara Lontz is a mathematician by both education (BA and MA in mathematics) and by profession. She has taught math to students in middle school, high school, and now teaches developmental mathematics to students entering Montgomery County Community College in Pennsylvania who are underprepared in her subject. Although her career path has taken her into the administrative position of Assistant Vice President for Academic Affairs, she continues to teach and plans eventually to return full time to the classroom, dedicating her efforts particularly to students who come to the college the least well-prepared in math.

"I realized," Barbara told me, "that we have been doing a very poor job at every level of helping students understand mathematics. I have decided that the most useful commitment I can make of my time is to help change that."

Barbara began by reviewing all of the major math education text series. "I must have 70 to 90 texts in my office," she said. "And I discovered that the approaches are largely the same. They move students through a series of topical lessons about fractions or decimals or percentages, with pages of practice problems and drill, and very little time committed to teaching broader mathematical concepts. I decided I needed to take a very different approach with my students if I planned to get better results. Then I read Robert and Ellen Kaplan's *Out of the Labyrinth* and it inspired me to think differently about how mathematics should be taught."

Out of the Labyrinth is the Kaplan's explanation of what they have learned about mathematics and the teaching of mathematics from what they call "Math Circles": small group classes in which they create mathematics from scratch with their students. Recounting the first class session with a group of five-year-olds, the Kaplans describe drawing a line on the board with zero at one end, and the number 1 at the other.

"Are there numbers between numbers," they ask? One student, who has just turned five and a half, volunteers that ½ is in between there somewhere, and the Kaplans draw it on the line near zero.

"It doesn't go there," another student offers. "It goes in the middle."

"Why?," they ask.

"Because that's what one half *means*!" she says, and a lively discussion is underway about what numbers *mean*. By the end of the first sessions, the students have discovered that there are "a kazillion" numbers between zero and one. During the next nine classes students identify ways to show where these numbers belong on the line, invent decimals to show that there are some numbers that can't be written as fractions and demonstrate, in the words of the Kaplans, that "the human potential for devising math, with pleasure, is as great as it is for creative play with one's native language: because mathematics *is* our other native language."[9]

So Barbara Lontz created a curriculum of her own to teach this language, one that spends the first few weeks focusing on helping students understand where math comes from. She discusses the contributions of ancient Chinese, Babylonian, Arab, Mayan, and Roman cultures and how each added something important to the way we understand the world of numbers. She then begins to expand these concepts to show students how the topic areas that have frustrated them during their earlier experiences with math are simply ways to express these concepts to solve various kinds of useful problems.

"Early in the semester we were talking about addition in its different forms," she explained as an example. "I wrote a simple algebraic equation on the board and immediately heard a couple of groans from the class behind me. When I asked what *that* was all about, the students said, 'this is algebra. We just don't *get* algebra.' So I explained that this was exactly what we had been looking at the day before—adding quantities to find one or more numbers that were still unknown to us. And that one of the cultures we had talked about had helped us design a set of symbols that allow us to write the problem in such a way that we can express it very simply. I could see the lights go on. It isn't a maze of rules and frustration anymore. They seem to get it."

In Professor Lontz' basic math courses, using the same final assessment she and her colleagues have used before, pass rates have climbed from 35–40 percent to 65–75 percent, with similar increases in subsequent classes. One of her students remarked, "I finally understand math. I will never be able to thank her enough for giving me the thrill of that."[10]

Testing as a Driver of Learning

There is a second failing in our attempt to teach students powerful mathematical ideas. Our assessment systems are still based on measuring computational skills—and that too, very narrowly. While serving in the role of program evaluator for a National Science Foundation grant, I sat in on a planning meeting of college mathematics faculty from two colleges in the region who were attempting to align what they were teaching in college algebra with what students had learned coming out of high school. One of the group members had created two Math Collaborative groups that brought together middle school and high school teachers from five school districts in the area to review curriculum and identify important areas were students were struggling. This professor noted that in conversation with these teachers she had mentioned that students were coming to the college math classes with no background in several key concept areas, including logarithms. She asked if the high school teachers could cover these subjects more completely in their Algebra II classes. The teachers responded with an emphatic "no!," explaining that the concepts she mentioned were not on the state assessment exams, and their administrations did not want them spending time on principles or concepts that would not be tested. The professor asked if the teachers thought this might be viewed differently if the colleges could help with training for faculty that would allow them to cover what students would need to know for state proficiency exams, and still master these other important skills.

"No," the teachers said. "We don't think our principals would allow that." The result was that the colleges backed their curriculum up to cover concepts that should have been covered in high school algebra, and had to reduce time on other math principles that had previously been part of the College Algebra course. Witzel and Riccomini in an article on maximizing the effectiveness of math education in the secondary schools note that this use of what they call "pacing guides" in middle and high schools to determine what will and will not be taught is not at all unusual and that the use of these guides "helps teachers eliminate any unnecessary concepts or skills not aligned with state standards or not included on end-of-year state assessments."[11]

Mastery Can Mean "Mastery"

Fortunately, there are math teachers in U.S. middle and high schools who are doing an exceptional job of teaching math to some of our most challenged populations and have figured out how to prepare students well for end-of-year testing *plus* master the concepts needed to be fully prepared for further math education. In a study of nine schools at both the middle school and high school levels that had been particularly successful with math education for students from poor socioeconomic backgrounds, the researchers found that "the most successful teachers do not limit themselves to simply preparing students for what is on the test. Instead...the most successful teachers range through the entire curriculum."[12] Like the teachers I saw at Delta Prep in Helena, Arkansas, these teachers realize that if students are given a solid and comprehensive foundation in mathematics, they will do well on the state math exams—even when the entire focus of instruction has not been on test preparation. A teacher at YES College Preparatory School in Houston, another charter school serving students from largely disadvantaged backgrounds, commented:

> Definitely [my teaching is influenced by the test, but also] it's influenced by standards quite a bit....But the way in which they are assessed [on the test] I think is a lot lower level thinking than what I

was doing today. So I try in my class in general to meet the standards, but not use those as a kind of minimum level of what I should be doing.[13]

Student performance in these schools indicates that the curriculum we have available can lead to levels of proficiency that are internationally competitive, though some of the teachers such as Barbara Lontz who are producing the best results choose to create their own curriculum as they go along by strengthening the concept that mathematics is one of our languages.

A study by the Institute of Education Sciences indicated that over 90 percent of elementary schools in the United States use one of seven commercially produced math curricula, and that some of these are more effective than others.[14] But teachers and teacher interest seem to make the greatest difference to student achievement, and a number who taught in the schools reviewed by Kitchen and his colleagues would agree with the middle school teacher who said, "every single person here that I know, supplements, supplements, supplements....I don't think we focus on any one text to do the job, because we don't....That is where we, as teachers, have needed to be much more creative and supplement that curriculum with some understanding and motivation and fun."[15]

Better Math Training for Teachers

The truth is, however, that many teaching mathematics in our schools may not have the background that prepares them to teach creatively and with understanding. Mahesh Sharma, provost of Cambridge College and professor of mathematics education for twenty-eight years, admits that many teachers in the United Stated are underprepared in the subject. He notes:

> The training of elementary school teachers includes a great deal of language arts and social sciences and very little mathematics and natural sciences. On average, they take no more than one undergraduate course in math. They themselves are afraid of math...when you hire a third-grade reading teacher, you don't hire a teacher who has a sixth-grade reading level. You want a teacher with a college reading level. So why do we hire math teachers who only know their grade?[16]

Sharma maintains that elementary school teachers should have at least three courses in math at the undergraduate level, and middle and high school teachers should have a degree in mathematics. Districts should then provide ongoing professional development to encourage math faculty to continue to be creative in the approaches they bring to the subject.[17]

A school that is in the midst of transformation may not have the luxury to dismiss all teachers who don't meet this requirement, and may not want to lose faculty who are talented teachers in other subject areas. But the transformation process should at least include a requirement for additional education for those with weak backgrounds in math, providing enticements to complete this training through a program that not only improves math skills, but helps teachers gain a personal appreciation for the language of mathematics. A teacher interviewed for the study of highly successful schools said:

> I'm really interested in, do they have the big picture? You know, why do you study these things? So whether it's my Algebra II class, we're doing augment inverse matrixes, I want them to know not just how do you go about solving the problem, but why you go about doing that. What led to the development of those ideas? Obviously I want them to be able to solve those problems. What I've seen over the last few years is that if they don't get the big picture in the beginning, all they're going to do is memorize their way through it. And that's not the kind of teacher I want to be.[18]

Great teachers talk about helping students learn how to "communicate in mathematics." Their elementary students are not only able to explain to each other how fractions, decimals, and percentages are all ways of understanding parts of a whole, but can explain why one approach to looking at a problem has advantages over another in certain circumstances. Students in upper level math courses not only know that logarithms are the exponent that indicates the power to which a base number is raised to produce a certain number and how to find that on a table, but why—and in what circumstances—that might be a useful thing to know.

Emphasizing Science and Technology

If we see learning the language of mathematics as opening the door to technology and the sciences, we also need to change the focus we are placing on these subjects in our schools, or our newly literate mathematicians will have no one to talk to. Since the boom in science interest that followed the Sputnik revolution in the late 1950s and 1960s, science education has again taken a backseat to English, reading, and even math. Students who are not following a specific college-bound track in high school can graduate in many states with no science beyond their sophomore year, leading experts to believe that we are losing our status as an international leader in the sciences.

Richard Florida, writing in the Harvard Business Review, maintains that we are facing a creativity crisis as a nation, in large part because our ability to generate innovative ideas is diminishing—or is being surpassed by other nations. Florida argues that "To stay innovative, America must continue to attract the world's sharpest minds. And to do that, it needs to invest in the further development of its creative sector. Because wherever creativity goes—and, by extension, wherever talent goes—innovation and economic growth are sure to follow."[19]

Creativity is not, of course, limited to the sciences, but in an age where advancing technology is driving economies, innovation in the sciences must be part of the creative package. During the past century, the United States depended heavily on international students who came here to study in science and engineering to fill many of the critical creative positions in business and research. Yale President Richard Levine noted that nearly half (43 percent) of the nation's Nobel Prize winners in science have been foreign born.[20] A high percentage of those who still fill our graduate programs in science and engineering come from other countries, and although graduate enrollments in these disciplines have grown steadily since the turn of the century, nearly 30 percent of these students are here on temporary visas.[21]

Many of these talented international students now choose to return to their home countries to work, and we are not replacing them with undergraduate talent from our own population. The United States has lost its international lead in the number of scientific papers published each year, and other countries and their innovators are rapidly passing us in the number of patents filed annually in our own country.[22] This shift in innovation and scientific preparation led Peter Drucker in an interview with Fortune Magazine to observe:

> The dominance of the U.S. is already over. What is emerging is a world economy of blocs represented by NAFTA, the European Union, ASEAN...India is becoming a powerhouse very fast. The medical school in New Delhi is now perhaps the best in the world. And the technical graduates of the Institute of Technology in Bangalore are as good as any in the world.[23]

What, then, must the United States do to remain competitive in the sciences and technology? A beginning point is that we must produce more mathematicians, scientists, and engineers, a task that will require the stimulation of greater scientific interest in the elementary and secondary grades. Though there has been general agreement that science education needs to be improved in our schools, there is an ongoing debate about how it should be done. But an evaluation of science education reform conducted shortly after the beginning of the new century indicated both a shift in method and some consensus about where science education should be going. When successful innovations in science and technology education were examined in thirteen OECD countries, it was evident that even in nations with highly centralized education systems, increased responsibility had been placed with capable teachers who were able to determine not only methods of instruction but also content. Innovative science education occurred when enthusiastic teachers, well trained in their disciplines, were given the latitude to determine how and what they would teach, when given specific learning objectives.[24]

The Teacher as Content Specialist

This emphasis on the teacher as judge of both pedagogy and content reinforces the importance of content-rich teacher preparation, and the necessity for those teaching in the sciences to have a degree in the discipline, complemented by training in teaching methods and classroom management. In disciplines such as the sciences and technology, this essential core of knowledge must be continuously refreshed and updated through an assertive program of professional development. Atkin and Black observe:

> such continuing education for teachers cannot be conceived solely as transmission of new knowledge and skills. Time and opportunity are needed to step back, to reflect, to consider how new ideas correspond with current practice, to hear from and share with colleagues who work with the same concerns and in similar constraints. If regular opportunities of this kind were common, it would be clearer that teaching is a profession that is taken seriously.[25]

Research is also suggesting that there needs to be greater integration, or at least "coordination," in science education, enabling students to see how biology relates to chemistry, and chemistry to physics. For too long, these disciplines have been presented as silos, as disciplines independent of each other. One of my brothers, a research chemist, explains that his area of specialization in high energy laser chemistry that involves blasting particles into their atomic components is as much physics as it is chemistry. Since he has been involved in making or redesigning some of the lasers he uses, his discipline is also deeply immersed in technology. Yet throughout our secondary education experience, these subjects are presented just as they are on the course schedule—as discreet blocks of information with little relationship to each other. Recognition of these relationships is completely absent if students are required to have only one laboratory science in high school, suggesting that, as with mathematics, our minimum expectation in the sciences should be at least two areas of study requiring laboratory work, with some effort to integrate these units.

We must also be careful that the emphasis we place on state or national assessments does not begin to define which sciences are important, and what content within each discipline deserves exclusive attention. Matthew Arnold, best known for his contributions to England's poetic literature, served as a school inspector in nineteenth century Britain and noted that a "payment by results" system introduced in the country that rewarded schools for student test performance would result in a narrowing of curriculum to the exclusion of critical subjects and content.[26] Two centuries later, our current emphasis on "funding for results" presents the same risks, unless we focus on assessment tools that measure the full range of learning in the sciences, or offer a curriculum of sufficient breadth that a student who has mastered its content will do well on any assessment.

In math and sciences education, three themes appeared in all of the highly successful schools—themes that have been part of this discussion of education reform from the opening chapter. Successful schools have high expectations of students, support their efforts to achieve, and accept no excuses for doing poorly. The curriculum is challenging, coordinated, and content-rich, with a focus on critical thinking and on mastering concepts as students move from grade to grade. Teachers work collaboratively, exchange ideas, and coordinate and align curriculum so that the colleague at the next level or in a related discipline knows that students will be prepared, and that catch-up work will not be necessary. Perhaps most importantly, students learn from the earliest grades that they are capable—that math and science can open doors for them that other disciplines will not, and that neither they nor anyone else should deny them access to those doors.

CHAPTER EIGHT

Second Language from the Start

If a little knowledge is dangerous,
where is the man who has so much as to be out of danger?
—Thomas Huxley

A few years ago there was a joke circulating about a distinguished, European-looking gentleman who walked into the lobby of a hotel in Average Town, U.S.A., and approached two clerks who were registering guests.

"*Parlez-vous Francais?*" he asked politely.

The clerks looked at each other in confusion.

"*Habla usted Espanol?*" he offered.

Still, nothing but blank looks from the clerks.

"*Sprechen sie Deutsch?*"

The clerks just shook their heads, and the man turned and walked out in frustration.

One of the clerks turned to the other and said, "Maybe we should learn another language."

"I don't see why," the second replied. "That guy spoke three and look how much good it did him!"

This story is humorous in part because we know it is so like us as Americans. Curt Hansman, an art historian who worked for several years in the National Palace Museum in Taiwan, relates that she often ran into Chinese who refused to believe she was American because she spoke fluent Chinese—and, in their experience, Americans didn't speak Chinese. While traveling with my younger sons in Brazil, one of whom speaks fluent Portuguese, we had a similar experience when a guide asked if we were Australian and, as we shook our heads, guessed Canadian, British, and New Zealanders. It didn't seem to occur to her that an American would speak fluent Portuguese.

Needless to say, we are not the world's most multilingual people, and we find the deficiency easier to excuse than to remedy, pointing out that so much of the rest of the world learns English. Some even speculate that, within two or three generations, English will be the universal language, or that there will be a device that can be attached to the ear like some Bluetooth or Star Trek gizmo and our speech will be translated into whatever language we choose. We present these arguments to spare us the embarrassment of trying to explain why we have never taken the time to learn to communicate with others in their own tongue, but have expected everyone else to learn ours.

Why Second Language?

There are a multitude of problems with the suppositions we put forward as excuses. The "English as the common language" argument is fading as we see greater numbers of international students choosing to learn Mandarin, because they suspect *it* may become the language of international commerce. A report in the Chronicle of Higher Education quoted a Korean girl who had chosen to attend school in Beijing as saying, "Sino-Korean business is really developing. Companies need people who can speak Chinese."[1] Our language's position of prominence in the past has been based largely upon two realities: the economic and media dominance of the United States in world markets, and the influences of British Imperialism during the eighteenth and nineteenth centuries in Africa, Asia, and the Americas. British imperialism is certainly a thing of the past, and many former colonies are doing their best to shrug off the last vestiges of foreign control, including placing new emphasis on indigenous languages.

Though we continue to be the world's dominant economy and generator of mass consumption pop culture, that position will certainly be challenged as China grows as a significant player in Asia, joining other influential economies along the Pacific Rim. Currently there are two and a half times as many people in the world who speak Mandarin as a first language, as those speaking English, and economists are now forecasting that China may be the world's leading economy by 2030![2]

The Arab world and the broader Middle East will also expand their influence during the first decades of the twenty-first century economically, politically, and strategically. Our experience in the Iraq and Afghan wars has provided dramatic illustration of the strategic challenges we create when few of our citizens speak Arabic or Farsi. And critical natural resource developments in South and Central America give Spanish and Portuguese greater, rather than lesser, importance, with Russian again joining French, Spanish, Arabic, Japanese, and Mandarin Chinese as languages critical to the strategic and economic well-being of our nation. Futurist Mark Milliron points out that if we are looking for the world's next superpower, patterned after the position held by the United States in the middle of the nineteenth century, we might very well look to Brazil—a vast country with a scattered population, rich in natural resources, and governed by an ineffective and occasionally corrupt government![3]

Even if English were to emerge as an international language, or if translation headsets eliminate all language barriers, we cannot expect our students to become comfortable global citizens without facility in other languages. Some years ago, the Director of the Long Beach Chamber of Commerce eloquently summed up the difficulties that our lack of facility in other languages creates:

> The world has become our main street, but the basic American insensitivity to other cultures prevails. It will continue to prevail as long as we continue to place the entire burden of language and cultural assimilation on our foreign friends. This was dangerous enough when the world was much larger and we were separated by great oceans. In recent years, it has led to monstrous errors and miscalculations in foreign policy, trade relations, and decision making within business and industry... The inability of the average American to communicate in the most elementary way in a second language continues to breed resentment in the minds of our friends in other countries. Today, and in tomorrow's world, the young man or woman of America who is fluent in a second language has a very bright future indeed.[4]

Ironically, this statement was made over a quarter century ago. The world is now much smaller and "flatter," our economic dominance has dwindled, and we have become much more multicultural as a nation. Yet the facility of most Americans with a second language has remained unchanged. Estimates of bilingualism in the United States range from 6 percent to 9 percent, with a large portion of these citizens being people who have grown up in bilingual homes. The World Watch Institute contrasts this with the rest of the world where they estimate that 66 percent of children are raised speaking more than one language.[5]

Ticket to Employability

The sad truth is that the mechanisms are in place to solve the problem— our schools. We simply choose not to take advantage of them. Virtually every school district in the United States teaches one or more languages other than English, but in most cases, aside from brief exposures for a few hours in grade school, language instruction as an option does not begin until junior high. Any broad-based academic reform in the United States must include increased and improved instruction in modern languages, and this is especially challenging if we are to rely on community-initiated change. To put it bluntly, in too many parts of our country, provincialism, nationalistic arrogance, and general lack of awareness about our changing position in the global economy blind us to the importance of learning other languages. But the need is real, and is growing. Purely from the standpoint of personal economics, Jerold Weatherford notes:

> A second language is now becoming a vital part of the basic preparation for an increasing number of careers. Even in those cases where

the knowledge of a second language does not help graduates obtain a first job, many report that their foreign language skills often enhance their mobility and improve their chances for promotion.[6]

A friend and colleague who served as president of a large college in Northern Illinois indicated that his institution had trouble keeping up with the demands of Chicago area industries with international needs. "I had occasion to give a speech where I mentioned that we had just taught Finnish to the work force of a company that had bought a company in Finland," he explained. "A gentleman came up to me after and said, 'we have just been acquired by a Finnish company and wondered if the college could teach us as well.'"[7]

In today's global marketplace, skill in a second language serves as a form of personal economic capital, in much the same way that being a highly talented three-point shooter in basketball or an all-state running back in football enhances personal economic capital.[8] Yet, according to the NCAA, the odds of a high school athlete in basketball becoming a professional in the NBA are roughly 550,000 to one, and in football in the NFL 980,000 to one. (Baseball is a little better at 455,000 to one!)[9] Yet hundreds of thousands of young Americans spend countless hours on the basketball court or ball field with dreams of becoming a professional athlete. If they were to spend the same amount of time learning fluent Chinese, Russian, Arabic, or Farsi, they would be virtually assured a productive career or livelihood. School districts encourage this misplaced set of priorities by preserving athletics at the expense of language programs when budgets become tight, and by sending performance reports to the local newspaper on a daily basis on each of their sports programs, while failing to submit a news release about the successes of the Spanish Club in regional competitions. If a school district *genuinely* wished to make each of its graduates employable, it would insure that every diploma was an indication that the recipient was fluent in a second language.

Language as Learning Training

Putting economics aside, the benefits of language study extend well beyond those offered by well-paying jobs and ease of communication. A study by Armstrong and Rogers randomly selected third graders from an elementary school and provided thirty minutes of Spanish instruction to one group, but not to the other. At the end of one semester, students were tested using the Metropolitan Achievement Test (MAT), and although there was no difference in reading ability, the students who had received Spanish instruction scored significantly higher in math and language aptitude.[10] In an earlier study of middle school students, researchers found that students who had received foreign language instruction had significantly higher performance in reading comprehension, language mechanics, and language expression in English than did their peers who otherwise demonstrated the same academic aptitudes.[11] This last finding makes some sense when we consider that most of us learn the structure of our own language simply by growing up with it. Few of us remember learning what the "subjunctive mood" is in English class ("If I *were* you"). We are either raised using it because we hear it at home, we don't learn to use it correctly at all, or a teacher in a French or Spanish class points out that there is *also* a subjunctive mood in English.

A series of studies conducted over half a century have also found that students of a second language become better students, in general. Those who had at least two years of second language in high school did significantly better on both the math and verbal sections of the SAT test, and on the English section of the ACT, even when the studies were controlled for basic intellectual ability. Students with second language experience demonstrate greater creativity, and seem to be more skilled at drawing inferences. There have also been studies that indicate that students of second languages simply become more skilled at "learning how to learn." A young person with a variety of experiences, even though they may be primarily intellectual, seems better able to develop the means to seek and assimilate new knowledge as it comes along.[12]

In its "College-Bound Seniors: Total Group Profile Report" for 2008, the College Board reported that students who averaged four or more years of foreign language study scored higher on all three sections of the Scholastic Aptitude Test (SAT) than did those who had studied four years of mathematics or natural sciences. There is undoubtedly a selection process that occurs as students from families where education is valued are encouraged by parents to "stick with foreign language study," but the SAT results demonstrate an exact correlation between years of foreign language study and test performance, even when students have only one or two years of language instruction.[13]

Through learning another language, we begin to examine the structure and organization of words and phrases and recognize the same features within our own. We realize that many of our expressions have roots in other languages, that practically everything we eat (aside from corn-on-the-cob, pumpkin pie, and grits) came from other cultures, and that we are, in fact, linked to the rest of the world already through speech and diet. The study of language is the study of culture, and we learn that we share holidays, fashion, humor, and music, as well as food. The more fully we understand another language, the more completely we appreciate the cultures that surround it.

My brother-in-law, Richard, is a language teacher whose observations about the value of language encouraged me to get back into the books and tapes and brush up on the language teaching of my own school years. Richard has always had a fascination with language itself, intrigued by the sounds he knew had meaning to someone else, but meant nothing to him. As a boy, he wanted to "break their codes" and understand those "secret" communications that no one else around him understood. He is one of those fortunate people who learn languages easily, and has mastered several. But he feels that the greatest value of language study wasn't apparent until he lived in and visited countries where the languages he knew were spoken.

"I found that barriers dropped immediately when I used, even imperfectly, the native language," he observed. "People saw it as an effort on my part to participate—as an indication of respect and appreciation for them and their culture. I was immediately welcome where others were still viewed with reserve and distance."

"I also learned," Richard told me, "that I could better see and feel a country from the inside-out, rather than from the outside-in. I feel a sense of vastness and reality in places where I understand and can communicate, rather than a sense of claustrophobia and distance where I can't."

I came to fully understand the latter experience when I joined a group of Vietnamese educators for dinner with two representatives from an Australian University. We were there to present proposals that would provide a newly developing Vietnamese university with a freshman and sophomore curriculum transferrable in total to a college in the United States or Australia. My presentation was well-developed, and the specifics of student's transition from Vietnam to the United States were clear and explained in detail. I was accompanied by an exceptionally skilled interpreter but still had one decided disadvantage. Both the Australians spoke Vietnamese. Shortly into the dinner I found that the conversation was happening all around me and included my interpreter, but I was only marginally involved. My Vietnamese hosts struggled awkwardly to keep me in the conversation, but I clearly wasn't—and I knew what Richard meant by feeling distant and claustrophobic. Needless to say, the contract went to the Australians who the Vietnamese saw as being better attuned to their needs, and more interested in them as a culture.

Language Instruction—Early and Often

Although still subject to considerable debate, much of the research on second language acquisition indicates that we learn languages most easily when we are young, and that there is a "critical period" during which we can hear, remember, and replicate sounds with greatest facility. (I even heard a news item on the radio recently saying that research is showing that babies cry with the accent of the language into which they are born!) Here again, there is much that we can learn from our international competitors in education.

A study of second language instructional practices in nineteen countries by the Center of Applied Linguistics found that seven had compulsory foreign language instruction by age eight, and eight additional countries required it before students finished elementary school. In a number of these nations, as students reach their upper grades, some of the content-specific curriculum, discussed in chapter 6, is taught in a second language—becoming "the medium of instruction in non-language subjects."[14]

Nations that lead the United States in academic achievement also take a different approach to the preparation and ongoing professional development of language teachers, an oft-cited factor in creation of excellent language programs. In countries such as Great Britain and the Netherlands, study and work abroad are expected parts of foreign language teacher preparation. Language instruction typically includes an undergraduate degree in the language, followed by a year of instruction in pedagogy and language teaching methodologies.[15] In many of our states, teachers can become certified to teach foreign languages with only a minor, or eighteen credits in the language, and with no particular evaluation of spoken proficiency.

With so few Americans speaking second languages, and even fewer fluent in critical tongues such as Chinese and Arabic, a concentrated focus on K-12 language instruction will require imaginative approaches to finding new teaching talent, providing even greater justification for liberalizing teacher certification requirements. The young Vietnamese woman who lived with my family was a trained language teacher, spoke excellent English, and would have been a capable teacher of Vietnamese—except for the certification to teach in an American school. Hundreds of young teachers in China, Russia, Egypt, or Saudi Arabia would jump at the opportunity to teach in the United States for several years. But districts would need much greater latitude in recruiting, hiring, and paying these faculty than what now exists.

There are wonderful examples in the country of public schools that are hiring teachers with international backgrounds and are providing effective language instruction, beginning at the earliest grades. My niece's daughter Marie, who attends Orchard Elementary School in Orem, Utah, entered an immersion program in Spanish as she started her second grade year and, as a fifth grader, now tells her mother that she is starting to "think in Spanish." Each grade at Orchard Elementary is divided into four classes, with one of the four sections taught completely in Spanish, with the exception of time spent on math and English. In nearby Layton, one of Marie's cousins has just begun kindergarten in a Chinese immersion program. In Marie's school, all of her teachers during the four years of immersion have been native Spanish speakers, each from a different Latin American country. In addition to language learning, Marie is immersed each day in discussions about these countries and their cultures. In order to participate in the immersion section, students must be achieving at grade level when they enter the second grade and parents must elect to enroll them in the program, but there are no other special requirements.[16]

A mile from Orchard Elementary, Rocky Mountain grade school has been declining in enrollment for the past decade as homes sold in the neighborhood have been repurchased by older couples, and as younger family's children move into middle and high school. Jennifer Cuevas' daughter Elayndria began in the Spanish immersion program at Rocky Mountain, but with the declining enrollments and budgetary constraints, several years ago two grades were combined to maintain an immersion section. Last year the program was moved to an "after school" option because classes had become too small to be maintained as part of the daily schedule. Students interested in Spanish immersion now meet three days a week after school for an hour of Spanish, with parents contributing $50 a year to maintain the program.

"The school still subsidizes it," Jennifer says of the program, "and we love it. No English is allowed. The class is kept fun and exciting and Elayndria's Spanish has become much better. She now carries on regular conversations in Spanish." According to Jennifer the junior high curriculum was changed to include advanced Spanish courses, and it is not uncommon for students to pass the Advanced Placement (AP) Spanish exam when they complete the ninth grade. "The schools don't always offer gifted programs," she said, "but use the language programs to meet that need. Since any child who is working at grade level can be involved, it is more accessible than the typical gifted program, but provides the same kind of enrichment."[17]

During a discussion with a university colleague, in which I was describing the successes of these elementary school immersion projects, he commented, "but look at where these school districts are! These communities are in the shadows of Brigham Young University that has one of the most comprehensive foreign language programs in the country. Half the people in those towns went on Mormon church missions to other countries and speak another language. Those people value language education." And that is exactly the point. Under the leadership of capable administrators with the support of progressive boards, this public school district is transforming second language education—just by making it a priority. It is not a matter of resources, or a reflection that these children are any more capable than children anywhere else in America. It is a matter of choice—a choice that is available to every school district with the will to make it.

Imagine the change that could occur within one generation if each elementary school in America offered one language immersion section—Spanish at Orchard, Chinese at Greenwood, and Arabic at Garden View. Within fifteen years a quarter of America would speak a second language fluently, and the rest of the population would be acutely aware of their shortcoming.

CHAPTER NINE

"Let Me Show You The World"

"Much learning does not teach understanding."
—Heraclitus

The difference between learning to gain *awareness* and learning to gain *insight* might appear largely semantic, but there is a subtle and important distinction. *Awareness* indicates an increase in general knowledge about, or experience with, something new. It is the kind of learning we expect every student to acquire if engaged with a creatively developed, content-rich curriculum. *Insight* enhances that knowledge by providing new ways of seeing and understanding based upon this new awareness and experience. Another personal anecdote will help illustrate the distinction.

Awareness Without Insight

I spent the summer of 1994 in Pakistan with fifteen other educators studying Islam and how elements of *Shariah*, Islamic Law, were being integrated with secular law in various parts of the country. For six weeks we traveled from one end of this fascinating land to the other, meeting with prominent social, governmental, religious, and business leaders in five regions of the country. The entire experience was an "awareness" builder from the moment we arrived in Islamabad until we left Karachi. Our travels took us to the shrine tombs of the great Sufi saints in Multan, from Peshawar through the Khyber Pass to the border with Afghanistan, and down into Baluchistan to the tribal city of Quetta. Each day provided new insights, and I remember one in particular that presented a lesson I have seen repeated over and over in years since.

Three especially memorable days were spent in the village of Kalam at the northern end of the Swat River Valley, the remote region of the Northwest Frontier Province that has been the center of much sectarian violence in recent years. Kalam nestles beside the river at the end of a precarious mountain road, near the 7000 foot level in the Himalayas. It has remained relatively untouched by outside influence, with most foreigners who visit the northern provinces choosing the more accessible villages along the Karakoram Highway that runs into China, or venturing farther west toward Gujarat. Westerners were still enough of a novelty in Kalam in 1994 that children gathered around those in our group with fair hair and blue eyes, also excited about experiencing something new.

Electricity had not yet reached Kalam—at least not by power line—and the Pakistan Tourist Development Corporation (PTDC) guesthouse in which we were staying offered an interesting assortment of amenities. My roommate Bill Wilson, a political science professor from St. Michael's College in Vermont, and I were amused to find that the small bathroom in our room also served as the shower stall, with a drain in the center of the floor. The sink had no plumbing and also emptied directly onto the floor. After using the sink or shower, we drew all of the water to the central drain with a long-handled squeegee and dried the rest of the room as well as possible with a towel. But the PTDC did have a generator that operated a few hours after dusk to extend light to the guesthouse. During these precious hours, a cluster of villagers gathered around a television set in the hotel's dining area to watch programming beamed in via satellite by STAR TV from Hong Kong. One of those hours was filled with reruns of the American soap, *Santa Barbara*.

I came to breakfast early one morning and attempted a conversation with one of the proprietor's sons, a handsome and assertive boy of ten or eleven with a halting command of English learned in the village school. After exchanging a few phrases from his memorized dialogues, he paused for a moment, then asked, "You are American?"

"Yes," I said.

"Do you know Santa Barbara?"

"Yes. Santa Barbara is a city in America."

"Do you know Kelly and Laken?"

"Kelly and Laken," I asked?

"Yes...from Santa Barbara."

As the halting conversation progressed I realized that, to this boy from northern Pakistan, America was simply a replication of his Upper Swat Valley region; a cluster of villages where families had lived for generations and everyone knew everyone else. It was also apparent that he believed that the miracle of television simply opened a window on the real life, day-to-day activities of Kelly and Laken in the village of Santa Barbara. Since I was American, Santa Barbara must be near where I lived—and I should know Kelly and Laken.

Though there was a touching lesson in his provincialism, the boy's question taught me a great deal more about technology and its uses. Marvelous electronic capabilities by themselves do not extend our capacities to see, do, or understand differently as long as we apply them within the same limited contexts that we have utilized in the past. Television for this boy was simply a visit to another village, one where he knew no one, but he assumed that I would, being from America. Unless those exposed to technology can envision new contexts and can explore them in previously unimagined ways, these tools offer only additional awareness, not greater understanding and insight. It is probably fair to say that, as we gain insights, they may not always be consistent with what the awarenesses of our past have led us to believe—with our own values assumptions and with those we want others to hold. In fact, insight might often be no more than realizing that others hold strikingly different points of view, believe in them as passionately as we do, and that these views merit our attention.

Gaining Insight

Another particularly interesting discussion during this same visit to Pakistan between the women in our group and three or four affluent Pakistani women illustrates the point. Our host for a dinner in the city of Lahore was an important civil servant and the Pakistani women, most of whom were attending the dinner as spouses of other dignitaries, had impressive Western educations, and had traveled extensively. They were not the burka-clad women we had seen in some of the tribal villages, or the peer-from-behind-screens women who peeked at us during some of the segregated dinners we had attended in the provinces. Yet, by custom, and to some degree by law, their roles in Muslim society were limited, in the eyes of the women with whom I was traveling. Several of my American colleagues found an opportunity during the evening to let the Pakistani women know how sympathetic they were to the limitations under which they were forced to live, and wondered what could be done to extend greater opportunity to women in the country.

One of the friends of our host, an attractive, articulate woman who had studied at a major eastern university in the United States, smiled at the questions and asked:

"Why are you assuming that we would like our lives to be different?"

"Well, consider the opportunities that aren't available to you," one of our group said. "You must feel stifled and repressed."

"Are you completely contented with your own lives?" the Pakistani woman wanted to know. "My experience in America was that many women are not particularly happy with their circumstances there and feel stifled and repressed."

"Well, of course we're not *completely* happy. There are always some things each of us would want to be different. But I can choose to be and do what I wish—an opportunity that isn't available to you."

"Can you choose to live in a matriarchal home, where your mother and grandmother, aunts and sisters have a closeness and control of home life that I don't believe I ever saw in America?" The woman asked. "Can you wake up in the morning feeling free from the pressures of proving yourself on yet another day? One of the things that has always amused me about Americans is that they believe they know what everyone else needs to be happy and content," she said politely. "Perhaps we are the ones who should decide what happiness and contentment mean for us."

"It's the ability to decide, that we think you should have," my colleague said defensively.

"We all have exercised that ability," the woman replied. "Each of us could have stayed in America or England, but it would have meant giving up other things that are more important to us—just as you have had to decide what to keep and what to give up. But please don't feel that you have to decide for us."

This discussion about "what is best for people" could be extended to include discussion about what differences in attitudes might exist between those who live in collectivists societies and those who favor individualism as a social and cultural norm, or to a debate on the distinction between contentment by resignation or contentment by choice. In fact, it is exactly the kind of discussion that forces the questions that lead to insight.

A group of students who spent a semester studying in Canterbury, England, remarked that one of the most memorable lessons learned from the experience was that it is possible to live without an automobile.

"There, people walk everywhere," one of the students commented. "And even though they have a diet that you would think would lead to weight problems—bread, meat, and potatoes—practically no one is overweight. I had never understood the relationship between weight and regular exercise quite as clearly as I did after walking two miles each way to school every day. It became part of my routine and I didn't even think about it being inconvenient after awhile. We're really spoiled in that respect—to our own detriment."

I am not at all certain where insight comes from—what happens in our heads to connect previously unconnected bits of information in creative new ways. But I am convinced that it happens more often when we are active and diverse learners, and when our experiences in learning cross cultural and international boundaries. At about the age of the young Pakistan boy in Kalam, I approached my father, who was an English professor with specializations in Elizabethan and Victorian literature, to let him know I had just read Shakespeare for the first time in school and didn't see anything especially great about his writing. My father assured me that this wasn't surprising given my brief encounter, and instructed me to read and memorize passages from a dozen of Shakespeare's major works until I could quote them "as fluidly and fluently as Shakespeare intended them to be said. Then," he said, "you will begin to understand the beauty and poetry of the language." After a pause, he added, "You see, appreciation is largely a matter of exposure."

How We Learn

Two of the early and most influential contributors to learning theory as we have applied it in American schools strongly supported this particular perspective. Jean Piaget, whose life spanned most of the twentieth century, was a Swiss biologist-turned-psychologist whose "constructivist" approach to learning maintained, very simply put, that our experiences as children form a scheme or framework of understanding into which we fit later experiences to give them meaning. We gain knowledge through the processes of assimilation and accommodation—taking in the information from our experience and accommodating it to the framework we have created. The broader the experiences of our formative years, the more complex and accommodating our framework becomes, allowing us to give meaning to a wider variety of experiences as we grow older. When we have experiences or encounter information that does not seem congruent with our established scheme, we are inclined to ignore it, to reshape the information to fit our framework, or, if the weight of conflicting evidence becomes too great, reluctantly modify the framework.[1] Sometimes, even when the weight of evidence is overwhelming, people choose to reject evidence because it would require restructuring a schema they do not wish to abandon—as in the case of those who reject the reality of the Holocaust.

Piaget was an outspoken proponent of both content and context in learning, arguing that content gains meaning only to the extent that the learner can fit it into an understood context. Some strict constructivists argue that, based upon the critical importance of learning within context, memorization of a body of knowledge is not useful—that if students are unable to give meaning to this memorized content, it cannot be converted into lasting knowledge. But this seems to discount where the initial framework comes from and how it expands, if not through the continuous filing of specific information into related categories.

Piaget's American contemporary, John Dewey, espoused a philosophy often labeled Pragmatism, though Dewey preferred Instrumentalism. For Dewey, learning occurred most successfully when a continuity of experiences (one learning experience building upon the next) "interacted" with a learning situation in which what was being learned made sense in the present context. Ideas were "instruments" to be tested in each problematic situation for their effectiveness and applicability, essentially for their utility. Those tested ideas that prove to be useful are retained and add to our sense of what is true about the world.[2]

I have occasionally wondered how Dewey would account for my sons' fascination with trivia, and for their endless testing of each other over who remembers the most insignificant and seemingly useless facts about a film or television program. As Jeanette Shaw, editor of *The World's Greatest Book of Useless Information*, put it, "For some reason, our brains are always starving for the kind of trivia that will have no meaning to your everyday life."[3] Then I remember the observation of Sir Francis Bacon that "studies serve for delight, for ornaments, and for ability,"[4] and realize that this useless information also has utility if it gives us delight and serves as an ornament. If that is the case, however, we lose a great deal of delight and ornamentation in our society when our studies and experiences are too limited.

The learning theories of both Piaget and Dewey support the need for early relevant exposure by students to the broadest array of experience and ideas—in the one case to construct a scheme of understanding that can assimilate a variety of future experiences and ideas, and in the other case to craft a set of proven "instruments" or ideas that give meaning to, and allow us to, respond appropriately to a widely diverse set of circumstances.

Sociologist Ruth Fulton Benedict wrote that "From the moment of his birth the customs into which [an individual] is born shape his experience and behavior. By the time he can talk, he is the little creature of his culture."[5] We are each born into a packaging plant that immediately begins to close us into a box. Depending upon the circumstances of our birth and family and the breadth of our education, that box remains small or grows larger. Insight develops only to the degree that we are able to assimilate diverse knowledge and experience within the box into which we are packaged, or the degree that, in the words to the popular cliché, we are able to think or see outside of the box—to force ourselves to experience new contexts and new ways of analyzing information.

Thinking "Around the World"

The nine-dot exercise below is often used to illustrate how easily we self-impose limits on ourselves. It simply invites the participant to connect nine dots with four straight lines without lifting the pencil from the page or tracing back over a line. If you have seen it done, the solution is simple. If not, it is often an impossible struggle because of the assumptions made before one begins.

Why is the puzzle so difficult for a first-timer? Because there is an immediate assumption that the lines need to be contained within the imaginary "box" created by the dots. Nothing implies that in the directions. It is simply assumed by almost everyone who attempts the puzzle. The solution requires that the lines be extended *beyond* the box. To solve the puzzle, start at the bottom left and go up through the left vertical line of dots to a point *above* the box where another straight line can be drawn down diagonally through the top middle dot and the center dot on the right vertical row. If this line is extended until it is even with the bottom row, a straight line can come back across the bottom row of dots, completing a triangle, and the fourth line can then be drawn diagonally up to the right, connecting the last dots. But you have to be willing to go outside of the box to complete the puzzle, and until instructed, most people will not.

An interesting variation on the exercise is to try to connect the dots with only *three* straight lines! Realizing that we are in the process of removing limits, take another look at the nine dots and think of ways in which this might be done.

Got it? To do this, one has to accept that no matter how small the dots, they have dimension. They have a top, a center, and a bottom. If you begin a line on the top of the top left dot and extend it through the center of the top middle dot, across the bottom of the top right dot, it will eventually meet a line drawn through the center row of dots, slanting the other direction through the thickness of the middle row of dots. This center line will eventually meet at its other end, a line slanting upward through the bottom row from right to left. All of the dots are connected with only three straight lines.

A student once pointed out to me that if we were willing to view a straight line as bending, as it circles the globe, he could connect the dots with *one* line! Using the same slanting technique, he could run the line through the top row and off around the world until it came back through the center row, then around again to pass through the bottom row. He does "bend" the rules a bit, but illustrates them in another way. Sometimes the best way to see the new contexts that allow us to think outside of the box is to extend our thinking around the world. If we are to be a globally connected society, shouldn't our students have some idea where Sudan is, for example? Or Sri Lanka, Qatar, Senegal, or Uzbekistan? Or for that matter, shouldn't they be able to locate Columbia, Finland, or Morocco on a map? I have read arguments by educators that "identifying places" is not critical to sound geographic knowledge—that the critical issue is having a general appreciation for diversity and cultural differences. But I would still like the mail sorters at the post office to know that Guyana, Ghana, and Guinea are different places and on different continents, and so should every globally aware citizen.

There is something of a conundrum in saying, on one hand, that each of us, through a combination of our innate abilities, living circumstances, and education, is "packaged" within a box, but needs to be able to see beyond the packaging if we are to acquire greater insight. The answer, it seems to me, must lie in creating within the packaging both permission and opportunity to peel the wrapping aside and look outside. For many students, this will not be part of the packaging of their home environment, and must take place at school.

"Don't Know Much About History"

When my youngest son, Paul, was fourteen, I included him in a trip to Thailand. I was accompanying a jeweler friend who was interested in attending a gem show in Bangkok and in seeing if he could buy rubies and sapphires coming across the border into western Thailand from

Myanmar. As Paul and I studied the map to find the cities we wanted to visit in Thailand, his attention was drawn to the country's proximity to Vietnam.

"The whole Vietnam thing was about this little country?" he asked.

"Well, look where it is," I pointed out, giving him a quick five minute lesson on the 1950s mentality about the spread of communism. "There was concern about spillover into Laos and Cambodia, through those countries into Thailand and Burma, then down into Malaysia, Indonesia, and the rest of South Asia. This philosophy was referred to as the 'domino effect.'"

He scanned the map for a few moments.

"I don't even see Burma on this map," he said, unaware, as a freshman in high school, that Myanmar had once been something else—or that Myanmar even existed.

We talked about the long common border Vietnam had with its neighbors, and the problems this border had created for American military strategists during the Vietnam conflict when the U.S. government had pledged to stay out of Laos and Cambodia. Our heavy bombing of those border areas was still leading to major injury and loss of life as people encountered unexploded bombs and land mines dropped during the Vietnam War.

We discussed the Golden Triangle where Laos, Myanmar, and Thailand meet, creating a haven of mountainous jungle where drugs, guns, and other contraband can slip easily across borders and between countries. I pointed out Bangkok's ideal setting on the Gulf of Thailand, and how this position was contributing to its development as a center for commerce and economic growth in the new south central Asia. As we talked, I couldn't help but wonder how a young man could complete nine years of formal schooling without having discussed at least some of these issues in class.

The truth is that, in the state in which we lived, Paul was easily able to finish twelve years of formal education without taking a geography course. The periodic National Geographic Roper studies of geographic knowledge show that, as a result, our 18- to 25-year-old citizens have an embarrassingly poor geographic understanding of both the rest of the world and of their own country.

In the 2006 study, six out of ten young Americans (63 percent) could not find Iraq or Saudi Arabia on a map of the Middle East, while 75 percent could not locate Iran and nearly nine out of ten (88 percent) were unable to identify Afghanistan—all at a time when thousands of their peers were engaged in a war in that part of the world that was supposedly of some importance to the security of the United States. Over half who responded to the survey did not know that Rwanda and Sudan are in Africa, and 70 percent could not find North Korea on a world map. In terms of our own geography, half were unable to find the states of New York or Mississippi, shortly after hurricane Katrina had devastated Mississippi's Gulf Coast and the country was in an uproar over poor services to the survivors.[6] This lack of geographic knowledge is in large part a reflection of a sense that this is not important learning. Of those responding to the survey, only 28 percent thought it necessary to know where countries talked about in the news were located. But, as with language, the importance of geography in the curriculum for the schools our children attend will be directly proportional to parental interest. If parents insist that geography be offered and that it be taught well, it will be.

After language and geography, the third leg of the stool of knowledge essential to global understanding is history—and not just U.S. History or Western Civilization. As noted in chapter 4, without knowing something about Confucian philosophy it is difficult to understand what motivates children in China and other parts of Asia to achieve. To be able to appreciate the forces that drive Japanese business, it is helpful to study the mythological origins of the Japanese and their sense of divine destiny. African history and colonialism have shaped not only that continent but our own. And Islam has to become something more in the minds of students than simply a group of fanatics who blew up the World Trade Center—just as Christianity is more than the Inquisition or the Crusades. Islam is the single-most important influence shaping a sizable portion of the global population, and must be understood in its historical context to appreciate what it means to be Sunni or Shi'ite, and what difference that makes to cultural, political, and social perspectives that affect our nation and future.

Yet we have done almost as poorly with history education as we have with geography and language study. A National Assessment of Educational Progress conducted as we approached the close of the twentieth century indicated that only four out of ten high school seniors had even a rudimentary knowledge of U.S. History, let along world history. Only 1 percent of those taking the test demonstrated that they had the ability to analyze historical trends and interpret their impact. To quote the then education secretary Richard Riley, commenting on the test data, "...as the song says, students don't know much about history."

As our students study language, history, and geography to provide the "context" that Dewey, Piaget, and others have found so critical to finding broad meaning and understanding in their learning, their studies must also be imbued with rich discussions of music, literature, and culture, with attention given to how each of these elements relates and contributes to what students are reading, listening to, or thinking about. Ideally a skilled English teacher will have students develop written and oral communication skills around a solid reading list that includes the world's best literature, allowing them to understand that the Indian epic, the Baghavad Gita, raises important questions about race and class, and that the Chinese poetry of Tu Fu speaks to human dilemmas that still plague every teen.

Where Do We Provide Global Context?

If we want our children or grandchildren to have the comprehensive, glob-ally integrated learning that must be part of an internationally competitive education, it seems to me that we have three choices: put our children in private schools where they might get a more focused and rigorous educa-tion, homeschool them ourselves, or demand world class results from the public schools. Though I have no quarrel with parents who select the pri-vate option, it has not been my choice for three reasons. I am not convinced that, discounting the effects of selective student admission, increased paren-tal involvement, and the general influences of socioeconomic opportunity, private schools provide any better academic experience than do our better public schools. The same language, history, and geography courses taught by equally talented and dedicated teachers can and should be available to our children in the public setting. We simply have to insist as a collective group of parents that these courses be offered, that our students enroll in them, and that they approach their learning seriously. I attended both public and parochial high schools and found that class selection, parental support, and personal commitment were much more important to the edu-cational growth of my classmates than was school type.

I also reject the private school option because I know that, realisti-cally, it isn't available to all parents who want an internationally competi-tive education for their children. Many of my more conservative friends would point out that this is the perfect argument for vouchers—for com-plete school choice. But two elements of my personal background make me extremely leery of vouchers. The one value I *did* find in parochial education was that the schools were able to incorporate other elements of personal, social, and religious development into my education without restriction. We began each morning with a mandatory chapel service, and I was required to take religious instruction as part of the curriculum. But as an educator in the public sector for over a quarter century, I know the Golden Rule of public support—he who has the gold makes the rules. Public vouchers will be followed by public restrictions—limitations on admissions requirements, insistence on special services, restrictions on denominational practice. Advocates will say that these controls will not develop if the "money comes to the families" rather than to the schools. That certainly has not been the case with federal financial aid at the col-lege level, and I fear that it would not be at any other level.

I also feel that I have a responsibility to support and attempt to strengthen a viable system of public education in America. We cannot afford to have a country that is half-educated, and private education never will be the answer for all. To the extent that the most able students with the strongest family support are drawn into private education, we add to the likelihood that we will create a permanent underclass. The expense of maintaining a large, educationally disadvantaged population far exceeds the costs of developing an effective public system of education, and I prefer to con-tribute my public dollars to the latter.

Homeschooling, though often done very well, presents some of the same challenges to our education system as a whole. It deprives public schools of talented students and the support of dedicated parents and, like the private school option, can eventually lead to a public system that serves the disadvantaged, underprepared, or poorly motivated. One of the reasons parents choose to homeschool is to keep their student out of situations where violence, drug use, or generally unacceptable behavior present constant threats to their children's well-being. All the more reason to insist as a parent that we take back control of our schools and insist on discipline, security, and rigor.

I recognize that there are also a host of moral, religious, and values issues that encourage parents to keep their children out of public schools. They choose not to have their students exposed to what they consider secular ideas, to the teaching of evolution, to what they view as an unacceptable approach to sex education, or a myriad of other issues. As a college administrator, I saw hundreds of students who had been homeschooled enter the college environment and found that, in terms of their moral and ideological development, they fared no better than did students who had been schooled in the public arena, guided by involved and concerned parents. In fact, the college students I see who have the best academic preparation come from families where learning is a continuous process at home and a supported process while at school. These families read, subscribe to magazines, attend cultural functions, travel together, and engage their children in an ongoing discussion about what values are important, and why. They are involved in their community schools, take interest in their children's learning, and expect their students to accept a good part of the responsibility for education. They believe that schools should provide good instruction, but that students have a responsibility to attend, be attentive, and meet the teacher more than half way. Most importantly, they have been given permission by their families to peel back the edges of their packaging—to peer beyond the box and gain new insights from what they see. Their families believe that truth can withstand the scrutiny of honest investigation, and that a system of values based on this critical review will better withstand the pressures life will place upon their children once they are on their own.

To aid schools with creating an interest in global awareness, communities can take a lesson from Middletown, Ohio, where since 1981 the city has sponsored an annual Middfest International, featuring a different country each year. During a four-day celebration, Middletown exposes its residents to the history, food, music, and general culture of the featured nation, presented by representatives from the country itself. Middfest's Web site announces that:

> The main focus of Middfest is education which reaches all age and socio-economic groups. Lectures, performances and exhibits are provided through the year to highlight and explain specific facets of that year's selected country.

The week prior to Middfest's October weekend celebration is set aside especially for area schoolchildren to tour art and cultural exhibits on display inside the Middletown City Building and view performances by visiting singers, dancers and storytellers. All children are invited to participate in **Super Saturday**, a special offering designed for youngsters to view exhibits and see performances on the Saturday morning of Middfest weekend prior to the event's opening to the general public. **Youth Park** is open throughout the weekend with special children's activities and performances.[7] (Bold type appears in cited quote)

If education is to become both a community priority and responsibility, the Middletown model points each community in the right direction. But realizing that every town in America will not emulate Middletown, families should accept the responsibility at the most basic societal level. The simple survey listed below will help parents determine how successfully they and their schools are preparing students to survive and succeed as a global citizen. If you can honestly answer "yes" to these questions, you and your school may not need to be rescued. If not, you have some work to do.

My school system:
Offers several modern language choices, including
  non-European languages, beginning in
  the elementary grades.   Yes   No
Requires geography, of all students, for graduation.   Yes   No
Requires world history or civilization for graduation. Yes   No
Sponsors events to expose students to other cultural
  groups in the school and community.   Yes   No
Actively encourages involvement and support by
  parents/guardians for each child's learning.   Yes   No

As a parent, I:
Expect my child to take another language.   Yes   No
Insist that my child take geography, if available.   Yes   No
Insist that my child take World History or World
  Civilization, if available.   Yes   No
Insist that my school system introduce these courses,
  if they are not available.   Yes   No
Actively get involved in my child's learning by
  expecting high performance and responsible
  behavior.   Yes   No

At home, I:
Take an active interest in how my child is doing
  in school.   Yes   No

Expect my child to accept shared responsibility for what is learned at school.	Yes	No
Provide a rich learning environment through books, magazines, and appropriate computer access.	Yes	No
Take my child to school and community activities that will broaden his/her experience with new cultures and ideas.	Yes	No
Discuss ideas, places, and events that acquaint my child with what is happening in the rest of the world in a positive and constructive way.	Yes	No

This is not one of those "If you scored 0–5, you are a horrible parent" quizzes we see in the Sunday newspaper supplement. Instead, it is an indicator of areas where parents could do more to create a competent global citizen of their child, and a more complete and appropriate curriculum in their local school. One of the universal dreams of all parents is for their children to have a more complete, satisfying, and rewarding life than the parents have had. At the risk of sounding overly simplistic, the answer for most of us is a fairly basic one—and one that costs only our time, energy, and dedication. It is a solid, broad-based education for the next generation. It is true that in many of our more disadvantaged neighborhoods the gap between the schools I am describing and the current state of education seems a mile wide. But it can be bridged *without* a huge infusion of money. It is happening in Washington DC under the iron-willed tutelage of a courageous chancellor and a committed mayor. It is happening in Helena, Arkansas, as more and more families with the poorest means elect to place their children in a public option that insists on excellence. It is happening in Leopold, Missouri, where a community has decided that education is going to be one of its distinguishing characteristics, and it is happening in El Paso, Texas, where a school board has made a commitment that the needs of children must be placed above the wants of adults. It can happen in every community in America, if, as a community, we choose to make it happen.

CHAPTER TEN

Hiring, Developing, and Evaluating for Excellence

For rigorous teachers seized my youth,
And purged its faith, and trimmed its fire,
Showed me the high, white star of Truth,
There bade me gaze, and there aspire.
—Matthew Arnold

The successes of teachers such as Jaime Escalante, celebrated in the 1988 film *Stand and Deliver*, and more recently Rafe Esquith of *Teach Like Your Hair Is On Fire* fame, demonstrate that talented, dedicated teachers working in even the worst of circumstances can produce remarkable results. Charles McClain, whose career in education progressed from K-12 teacher and administrator, to community college and university president, to state commissioner of higher education, later served as desegregation monitor during a major court-initiated effort to improve the failing Kansas City School District. During this period of oversight, McClain noted that one of the elementary schools in the district with student demographics very similar to other Kansas City schools had noticeably better performance on the court-approved student assessment evaluations. His personal review of factors that might account for these differences revealed that a dedicated principal had systematically eliminated weak teachers and replaced them with faculty with strong academic credentials and a passion for teaching. It is McClain's view that teaching ability is strongly correlated with the academic rigor and performance of the teacher's own education. He also affirms the mounting body of research that demonstrates that no amount of resource support, curriculum innovation, or administrative commitment compensates for the effects of good or bad teaching in the classroom. Great teachers produce great results in almost any environment, and bad teachers fail under even the best of circumstance.

Great Schools Require Great Teachers

Earlier I recounted my experience at two schools in Iran where Bess Byrom, Jane Boggs, Charles Brewster, and Pam Young "showed me the high, white star of Truth," and inspired an otherwise difficult-to-inspire teenager to be excited about learning while my older sister, with a different set of teachers in the same schools, found her years frustrating and tedious.

First Lady Michelle Obama, for whom improvement in education has become a personal mission, noted that "in a 21st-century global economy where jobs can be shipped to any place with an internet connection and children here in America will be competing with children around the world for the same jobs, a good education is no longer just one road to opportunity—it is the only road. And good teachers aren't just critical for the success of our students. They are the key to the success of our economy." Mrs. Obama adds "...it's not surprising that studies show that the single most important factor affecting students' achievement is the caliber of their teachers."[1]

Study after study demonstrates that great teachers, using the same materials with students of similar preparation, background, and ability, get significantly better results—and that teachers' talents, enthusiasm, and commitment have much greater bearing on this success than does any formal teacher certification. In a review of these studies by the Brookings Institute, researchers found that when student achievement was compared in classes taught by teachers who were traditionally certified and by uncertified teachers, there was much less difference in student success based on teacher certification than there was between the students taught by the best and worst teachers in each of these credentialing divisions.

In the Los Angeles Unified Public School District a study involving approximately 150,000 students divided teachers into three groups: those who were certified when hired, those who were uncertified but participating in an alternative certification program, and those who were uncertified and not participating in a certification program. The study found that:

> The difference between the 75th percentile teacher and the 50th percentile teacher for all three groups of teachers was roughly five times as large as the difference between the average certified teacher and the average uncertified teacher. The difference between the 25th percentile teacher and the 50th percentile teacher is also about five times as large. And those larger differences are evident even after adjusting for the obvious socioeconomic and educational factors that affect student performance.[2]

This remarkable finding indicates that achievement by those taught by our most successful teachers can be *ten times* greater than those taught by teachers who are the least successful! In addition to calling into question the value of formal certification, these studies reveal that the differences in student achievement based upon teacher performance are not simply significant, but are monumental! Yet despite the well-documented evidence that great teachers produce great results and poor teachers lead to poor student achievement, we continue to hire and retain poor faculty, and fail to effectively evaluate them once in the classroom.

Insuring Effective Teaching

The New Teacher Project, a program developed to select, train, and evaluate new teachers with nontraditional backgrounds, noted in a 2009 report that:

> you would be dismayed to discover that not only can no one tell you which teachers are most effective, they also cannot say which are the least effective or which fall in between. Were you to examine the district's teacher evaluation records yourself, you would find that, on paper, *almost every* teacher is a great teacher, even at schools where the chance of a student succeeding academically amounts to a coin toss, at best.
>
> In short, the school district would ask you to trust that it can provide your child a quality education, even though it cannot honestly tell you whether it is providing her a quality teacher. This is the reality for our public school districts nationwide. Put simply, they fail to distinguish great teaching from good, good from fair, and fair from poor. A teacher's effectiveness—the most important factor for schools in improving student achievement—is not measured, recorded, or used to inform decision-making in any meaningful way.[3]

Readers were stunned and angered when Steven Brill, writing in the *New Yorker*, drew national attention to what are called "Rubber Rooms" in the New York City public schools—places where teachers who the district is attempting to dismiss for incompetence or misconduct while away their days drawing full salary and benefits while the Department of Education struggles to dismiss them. Though reasonable due process is an expectation in any teacher dismissal, a quagmire of procedural hurdles created by the United Federation of Teachers (UFT) makes it practically impossible to fire a New York City school teacher, and Brill suggests that much of this procedural web is designed to keep any teacher, no matter how inept, employed. One elementary school principal accused the union president of being willing to protect "a dead body in the classroom." Approximately six hundred teachers crowd these Rubber Rooms (officially called Temporary Reassignment Centers), with an average tenure in the rooms of about three years, and arbitration and settlements taking from two to five years. The first case tried under a program called the Peer Intervention Program (PIP) was underway when Brill wrote his article, and it was expected to take a year to resolve, with a cost to the city estimated at about four hundred thousand dollars. After a judgment is rendered, history suggests that the city will still continue to bear some expense for the teacher since arbitrators have been reluctant to fully terminate any employee.[4]

The Rubber Room is a condition of the UFTs 166-page, single-spaced contract with the New York City Schools, a contract that also includes a provision that, as of the fall of 2009, another seven to eight hundred teachers who are not providing any services to the district will be added to the payroll. This group, referred to as the Absent Teacher Reserve, were released because their schools were closed or the number of classes in a school was reduced, and have either chosen not to seek other employment or have not been able to find new positions. By 2010, the reserve list was expected to cost the city as much as a hundred million dollars a year and, in total, the district would be employing nearly seventeen hundred idle teachers.[5]

Writing a few weeks later in the *New York Times*, Op-Ed columnist Nicholas Kristof praised the Democratic Party's battle for reform in health care and their decades-long fight against poverty, but chided the Party for turning a blind eye on one of poverty's greatest allies—failed schools. He noted that "cowed by teachers' unions, Democrats have too often resisted reform and stood by as generations of disadvantaged children have been cemented into an underclass by third-rate schools," and pointed to additional examples from the nation's other coast.[6]

In an investigative piece by the *Los Angeles Times* that examined every attempt over a fifteen-year period by a California school district to fire a tenured teacher (159 in all), the reporter found that the process was "so time-consuming, costly and draining for principals and administrators that many say they don't make the effort except in the most egregious cases."[7] The review also revealed that even when attempts were made to dismiss the most blatant cases of misconduct, a third were overturned by a review panel that has the discretion to restore jobs even when grounds for dismissal have been established. Teachers who are evaluated as incompetent are often "coached" and remediated for years, leading one principal to observe that "an ineffective teacher can instruct 125 to 260 students a year—up to 1300 in the five years... it often takes to remove a tenured employee."[8]

We find ourselves facing a situation in our country where we know that education must radically be reformed if we are to remain globally competitive, where virtually all research on the topic demonstrates that good teachers are the most important contributor to student success, and where we have developed a web of protection for incompetent, unproductive, and malfeasant teachers that is costing us hundreds of millions nationally, while keeping bottom-quartile teachers in front of our classrooms. Most of the recommendations for reform covered in previous chapters—more rigorous curriculum, longer school days and years, earlier emphasis on foreign language—can be implemented by most districts without the support of new chartering privileges. But without greater ability to control resistance from these same teachers' unions, some will be difficult, and making essential changes in teacher quality and performance will be near to impossible. Schools must be extended the authority to select those best suited to teach in their classrooms without the restrictions imposed by certification and by union contracts. They must then be able to evaluate faculty based on student performance, and dismiss teachers who do not meet performance expectations, without unreasonable cost and labyrinthine procedural barriers. They must be free to eliminate tenure, and to require continuing education for faculty in order to remain employed. After noting that the results of grant-making efforts to reform school performance had been disappointing, Bill Gates noted in a report to the Bill and Melinda Gates Foundation that "A model that depends on great teaching can't be replicated by schools that can't attract and develop great teachers."[9]

Attracting and Developing Great Teachers

How then does a district attract and develop great teachers? Returning to the recommendation of the *A Nation at Risk* study, the key to "breaking through" for a community that wishes to have the best teaching staff in its region is to determine what it will take to transform its district into a globally competitive school system, then commit to pay its teachers enough to rise to that standard. Admittedly, there are districts like Leopold, Ysleta, and the Alpine District in Central Utah that produce impressive student results without higher teacher wages, but in many ways these are anomalies, and with the changes we need to make in teacher preparation and continuing education, we will not attract the quality we need without making the profession more competitive financially.

The OECD country that has made the most dramatic strides in education in the past fifteen years may be Korea, where the country has shown the greatest increases in both high school and university attainment, and now ranks first in high school completion for 15–34-year-olds, and second in college degree attainment. It is not coincidental that it has also been among the most aggressive in making adjustments in teacher compensation, now ranking second only to Luxembourg in teacher pay, and providing the greatest potential increases from the time a new teacher is hired, through fifteen years in the profession. Korea ranks fourth in average pay for a fifteen-year veteran, behind Luxembourg, Switzerland, and Germany, with the United States a distant twelfth. Teachers at the top of the scale in Korea average over twice the pay of their American counterparts! Two observations should be important here for American educators and policy-makers. Our international ranking in education attainment corresponds very closely to our ranking in teacher compensation, and as a nation that prides itself in being the world's wealthiest and world economic leader, we should be ashamed that teachers in England, Spain, Scotland, Ireland, and seven other countries are better compensated than are ours.[10] One of the notable differences between these countries and ours is that pay scales are used to attract some of the top graduates from the best universities into teaching, while in the United States we have chosen to make it one of the least well-compensated professions. As a result, those who enter teacher education programs often are not our best students, and are not graduating from our best universities. They lack the academic credentials of their international peers, and are weaker in content knowledge. In Finland, the country that leads the developed world in student achievement, only one applicant in ten is accepted into teacher education programs,[11] while in our own country teacher education is open to students with very modest high school preparation. It should not be surprising that, in a relatively short period of time, the dozen countries that place greater value on teacher preparation and compensation have passed us in student achievement.

As part of its commitment to create a culture of education in its community, CORE Teams and school boards need to make the case to the public that in order to enact the changes necessary to develop internationally competitive schools—longer days, longer years, weekend commitments, and better training—the community must be willing to pay its teachers accordingly. Then the board and administration can say to the faculty, "We are planning to pay you more than the leading salary schedule in our region, but this is what we will be asking of you in return. There will be no tenure in this district. Each teacher will be on a two-year contract that is evaluated and extended annually, based on student performance in the classroom. The school day will be one period longer, and the year will be extended by 15 days. There will be on-going professional development requirements that will be directed primarily at strengthening your content knowledge and global awareness. We will be asking parents or guardians to sign a commitment of support, and will be supporting you in maintaining high standards of discipline and behavior in our classrooms. You can decide if this is the kind of district in which you want to teach and if you believe you are equal to these expectations."

Why is eliminating tenure required on this list? Tenure was originally created to protect teachers from political patronage, from capricious decisions by board members who were upset with a teacher's point of view or selection of teaching materials, or from a vindictive district resident when a faculty member scolded her child. But all of these protections can be guaranteed by other contractual assurances of due process, without the accompanying guarantee of lifetime employment that underpins tenure. Tenure's principle value for teachers in the K-12 setting has become employment protection in cases of poor performance. No profession beyond teaching provides such ironclad guarantees, yet few employment protections can lead to such lasting damage. Teaching must be like every other profession in which performance expectations are clearly spelled out, evaluation is performance-based, and rewards and continued employment depend on meeting achievement goals.

At a Washington DC conference held by the Strategic Management of Human Capital project, experts agreed that what has been missing from schools "are broad, thoughtful strategies that link the major components of school districts' personnel systems—recruitment, hiring, placement, induction, professional development, evaluation, compensation, and termination—to their bottom-line goals for students."[12] In other words, student achievement goals should drive these processes, not be a byproduct of them.

Recruiting, Hiring, and Evaluating

In recruiting and hiring, if student achievement is the driving goal and if curriculum is to be content-based, teachers must first be content specialists—even in the elementary grades. In countries that routinely outperform the United States in evaluation of student achievement, all teachers must first have a degree in a discipline, with training in pedagogy serving as a supplement, rather than as the core of the degree. The *Teach for America* (TFA) project in the United States has demonstrated how effective even novice teachers can be when recruited based on strong academic and leadership abilities, then provided with a supportive professional development regimen. Candidates for *Teach for America* are screened based on "demonstrated leadership and achievement in academic, professional, extracurricular, or volunteer settings," and although some have degrees in education, many do not.[13] Even though these teachers have only a two-year commitment and the majority do not remain in teaching beyond their term of service, an independent study of their performance indicated that:

> TFA teachers are more effective, as measured by student exam performance, than traditional teachers. Moreover, they suggest that the TFA effect, at least in the grades and subjects investigated, exceeds the impact of additional years of experience, implying that TFA teachers are more effective than experienced secondary school teachers. The positive TFA results are robust across subject areas, but are particularly strong for math and science classes.[14]

A Louisiana study of teachers recruited and trained by The New Teacher Project, a similar initiative that attracts graduates from other majors and professions into teaching, found that recruits through the New Teacher Project outperformed experienced teachers in math and reading, and did as well in language arts and in the sciences.[15] These studies indicate that districts would be best served by having the freedom to hire graduates who have distinguished themselves as students and leaders in any discipline, then provide them with a robust internal training program the summer before beginning work and while on the job, patterned after the training regimen developed by these alternative teacher recruitment projects.

It is true that there are conflicting studies that indicate that teacher certification makes a positive difference in student achievement, such as Darling-Hammond and Associates' six year study of Houston teachers and student performance.[16] My interest here is not to support one approach to teacher training and preparation above another, but to illustrate that great teachers emerge from a variety of backgrounds, and schools should be given discretion to hire capable talent that has not passed through traditional training systems. Evidence does seem conclusive that good teachers, however prepared, are critical to student success. In essence, the nature of teacher preparation is less critical to obtaining positive results than are regular assessment and evaluation of those results and corrective action if teachers fail to perform.

Teach for America has developed a framework for ongoing teacher training that will be discussed in greater detail in the next chapter, but the program is very open about the basic criterion it used for teacher evaluation—student success. New teachers are expected to meet one of three standards during their two years in the program: moving students forward by one and one-half grade levels, closing performance gaps within their classes by at least 20 percent, or insuring that 80 percent of students meet grade level standards.[17] Assessments are administered at regular intervals, with detailed feedback to each teacher based on classroom observation as well as test scores, providing recommendations for improving student performance based on strengthening of five leadership traits the program has found relate directly to teacher effectiveness.

No area within education is more hotly debated and none more vigorously resisted by teachers' unions than is performance-based evaluation. Speaking at a Washington conference on Strategic Management of Human Capital, the vice president and professional development director of a large teachers' union said "the union is open to using data on students' academic performance as a way of guiding instruction and improving practice, but not for evaluating teachers."[18] Reflecting union influence on the political process in New York, in 2008 the state legislature prohibited student testing performance from being used in making faculty tenure decisions, requiring districts to extend this lifetime employment protection without taking into account whether a teacher's students were showing progress.[19] A California law passed in 2006 prevents student performance data from being used for teacher pay, evaluation, or tenure decisions.[20]

There is something so inherently absurd about assessing teacher performance without considering student progress that every parent in America should be completely incensed and should demand performance-based evaluation. Teacher assessments are now largely based on classroom management, presentational style, support of school activities and events, and coming to work every day. But in what other profession—particularly one that has such a profound and lasting effect on our own futures and that of our children—would we allow performance to be evaluated without considering whether the person does what he or she was hired to do? We would be infuriated if we knew that an airline was evaluating pilots without considering how skillfully they were getting passengers to their destinations, but we shrug it off when teachers don't get their students to the next grade successfully. We would insist that a doctor be stripped of her license if we knew that, despite having a great bedside manner, most of her patients came out of what was supposed to be routine surgery permanently disabled, but we don't even blink when our children complete school unable to read or compute. What *is* evaluation if not the process of determining whether an employee is effectively doing what he or she was hired to do—in this case, ensure students' academic success?

Value-Added Evaluation

Critics of performance-based evaluation point out that many other factors influence student achievement—home environment, previous instruction, basic intellectual ability. But this does not negate the reality that good teachers still get good results with students that poor teachers struggle to improve. In fairness, teachers working at the same grade level do get new students who are very differently prepared, among different schools and even within the same school. But if evaluation is based on the "value added" to the student's academic achievement compared to where the student was when she entered, school-to-school comparisons and teacher-to-teacher comparisons become of secondary interest. Using the *Teach for America* criteria, did the students progress one and a half grade levels from where they were when they entered? If most joined the class at grade level, did 80 percent leave ready for the next year? If the class is very diverse in terms of achievement, was the achievement gap narrowed by 20 percent? Other evaluative criteria that include whether the teacher manages the classroom well and isn't rude, abusive, late, or habitually absent should be considered as well. But the most important criteria must be student achievement and success, based on the value that was added to the student's education while in the class, as measured by some standardized assessment.

There is a basic assumption in "value-added" evaluation that the school knows the level of achievement for each student at the end of every year—an assumption that is probably not valid in most districts across America. States assess at key grade levels, or stage evaluations after every other year. To be fair to teachers who are being evaluated based on increases in student learning, an assessment system needs to be in place that evaluates critical learning objectives after each academic year. Faculty will complain that students already are "assessed to death," and the last thing schools need is more testing. But fair evaluation requires fair and regular assessment, and one advantage of an extended school year is that several days can be set aside at the end of each year for student progress evaluation.

A report by the New Teacher Project referred to our current approach to selecting and placing teachers as "the Widget Effect," reflecting the apparent attitude by schools that teachers are interchangeable parts that can be swapped out with no appreciable effect on students or student achievement. The report concludes that this attitude is symptomatic of schools' inability or unwillingness "to assess instructional performance accurately or to act on this information in meaningful ways."[21] After studying performance evaluation systems in twelve districts in four states, the study determined that, as a general rule, all teachers, with little attention to student performance, are rated "good" to "great." Outstanding performance goes unrecognized and poor performance remains unaddressed. No special attention is given to new teachers and their preparation, and professional development programs are generally inadequate. To reverse the Widget Effect, the study presents four recommendations as central to effective teacher evaluation, calling upon districts to:

Adopt a comprehensive performance evaluation system that fairly, accurately, and credibly differentiates teachers based on their effectiveness in promoting student achievement.

Train administrators and other evaluators in the teacher performance evaluation system and hold them accountable for using it effectively.

Integrate the performance evaluation system with critical human capital policies and functions such as teacher assignment, professional development, compensation, retention, and dismissal.

Adopt dismissal policies that provide lower-stakes options for ineffective teachers to exit the district and a system of due process that is fair but efficient.[22]

To provide greatest value, the system of evaluation must be part of every new teacher's orientation, and must be implemented during the first year of employment. *Teach for America* and The New Teacher Project demonstrate that new teachers can produce impressive results very quickly, and experienced personnel managers know that employees demonstrate, within a year, whether they have promise. After a rigorous selection process based on strong academic credentials, supportive references, and an impressive mini-teaching demonstration, the first few years of a teacher's contract should be consistently evaluated, heavily supported, but should remain probationary—freeing the district to release the teacher at the end of any year if performance is substandard and there is no evidence that additional training or mentoring will lead to significant improvement. To make this possible, student assessment must be an annual event for every grade level. Otherwise, a probationary teacher can be several years into a contract without any fair basis for evaluation, putting both the faculty member and the district in a difficult position if performance is suddenly determined to be unsatisfactory. Teachers who are performing poorly deserve frequent and evidence-based evaluations, as do effective teachers.

Following an appropriate probationary period, tenure should be replaced with a rolling, annually reviewed, two-year contract that allows a district, if deficiencies appear or teachers find themselves with a particularly difficult class, to provide another year of evaluation and assistance before a final determination is made about continuation. But the guiding consideration must always be student success. Two years of subpar achievement may leave students far enough behind that even the most talented teachers in their following grades will have difficulty bringing them back up to grade level.

Releasing with Sensitivity

When performance fails to meet expectations, dismissing an employee is difficult under any circumstance, and using a new performance-based system to let a longtime teacher go, who has been shown to be ineffective, is probably most challenging. But it must be done, and it can be handled with compassion and concern for both parties.

Mary Follett, whose approaches to leadership and management were discussed briefly in chapter 5, preferred to refer to conflict within an organization as "difference," noting that, when conflicts arise, whether between individuals or between the individual and the organization, they are reflections of differences in perceived values. In many ways, conflict is useful in that it draws attention to areas where disagreement or misunderstanding exists and needs to be addressed, preferably by reaching some consensus about how the values can be harmonized. The first step to reaching this integrated solution, Follett maintained, is to "put your cards on the table, face the real issue, uncover the conflict, bring the whole thing into the open."[23] In the case of a school district, as a starting point this requires that student achievement expectations from grade to grade be clearly and explicitly stated in policy as an expressed value of the district—introducing another Fundamental Law of the Educational Universe. "*Goals must precede data* and must be understood as the reason for gathering performance information." The school *does not* collect student performance information principally to evaluate teachers, but to determine if it is meeting its broader student achievement goals. Faculty evaluation is only part of assessing student success.

When the district's performance goals and a given class's achievement do not coincide, the opportunity develops for sitting down with the teacher, laying the performance information on the table, and discussing why the two don't correspond. There may be circumstances that all agree are beyond the teacher's control, or conditions that with some assistance can be remedied. It may also be that the expectations of the district and the abilities, ambitions, or interests of the teacher are not compatible. This needs to be acknowledged and the teacher assisted in finding a better employment fit.

A word or two more about "goals preceding data." If teachers are to be evaluated fairly with a minimum of debate when goals are not met, it is critical that the district clearly outlines its learning expectations for each grade level, and base these expectations on a reasonable and measurable standard of achievement. Teachers should, in fact, insist that this be the case. If these learning expectations are not explicit, evaluation will inevitably be subjective and arguable, and this is not fair to either the employee or the district. Just as it is only fair to administrators that boards develop policy, then leave policy objectives to the administrators to achieve, it is only fair to teachers that, once learning objectives are established, they be given a reasonably broad latitude about how to achieve them. If entrusted as a teacher with meeting the state's learning objectives for Algebra I, but limited to a text and lesson plan requirement that the teacher believes limits her ability to meet those objectives, she deserves the opportunity to demonstrate that, using other methods, she can achieve the objectives. She also deserves the support of the district in addressing areas of weakness that she knows makes it difficult to effectively meet the objectives, given reasonable time and investment. However, it becomes part of that "cards on the table" discussion to mutually determine where the deficiencies lie, what is reasonable time and investment for remediation, and when the results of the activity need to appear in performance results. Much of this negotiation can be minimized through a continuous faculty development program that is based on annually identified needs and tailored to individual faculty improvement—the subject of chapter 11.

CHAPTER ELEVEN

Keeping Teachers Current, Enthusiastic, and Energized

"If I should not be learning now, when should I be?"
—Lacydes

While visiting Angor Wat in central Cambodia, my wife and I spent a leisurely afternoon in a sprawling market in the nearby city of Siem Reap. As we were inspecting a collection of carved wooden elephants in a crowded stall, we heard two young women speaking to each other in front of a display of silk scarves hanging on racks across the narrow alley. From their conversation and accents, we decided that they must be American coeds, and as they walked out into the corridor of the market, we joined them and asked where they were from.

"Holland," they said, still speaking English that sounded as if they had grown up in Omaha. As we walked and visited, we learned that they had just completed their bachelor's degrees in preparation for becoming middle school teachers, but had delayed their careers to travel in Southeast Asia. They were midway through their itinerary, and were spending a month in each of six countries. Their adventure had begun in Thailand, where they had traveled north into Laos, then over into Cambodia, and from there were headed into Vietnam, Malaysia, and Indonesia. They explained that it was common for college graduates in Europe to save for years in anticipation of traveling for a period of time after college graduation, spending from a month to a year becoming acquainted with other parts of the world.

We asked what value they expected to gain from their travels, and one of them shrugged almost matter-of-factly and said, "This is the world we are going to be teaching our students about. Don't you think it is important that we have some experience with it ourselves?"

What separates these new Dutch teachers from most of those who graduate from our colleges of education every year and enter the teaching profession? For one thing, they spoke a second language as fluently as their own. They also had degrees in one of the area of content they would be teaching, rather than a degree in the "process" of teaching. But, more importantly, they saw their formal education as only a piece of what would be required to be a great teacher, and recognized that, in a world as completely interconnected and interdependent as ours, learning needed to be continuous and global.

It struck me that, philosophically, if we thought of teacher preparation as endowing the novice with three attributes—solid grounding in an academic content area, an understanding of the global context in which their students would live and work, and a clear grasp of teaching methods and classroom management strategies—these students and their countries had chosen to emphasize the first two attributes, while we focus almost entirely on the third. We also make no consistent effort to insure that content and global context are ever acquired, but spend four years helping students master technique. For the Dutch teachers, teaching was first about knowing. For us, it is more about doing. As part of our community-based effort to reform education, there is a perfect opportunity to reform the way we develop our teaching talent by shifting our emphasis from "how to teach" to "what to teach and why," then insuring that it is taught well.

Assuming that we can agree that a content-rich curriculum can best be presented by teachers who are content-specialists, this point of agreement indicates where to begin with teacher development: by turning each faculty member into a content-specialist. As products of our current system of teacher education, many are not. Granted, they may have areas of concentration or minors in a discipline, but by international standards this hardly qualifies them as specialists. Each of them needs at least a bachelor's degree in an academic discipline, even for teaching in the elementary grades.

Discipline-based Teaching Certification

Reporting on research that evaluated the effectiveness of teacher education programs, a panel convened by the American Education Research Association found that teachers who had completed five-year programs requiring a major in a content area in addition to their teaching credential rated themselves higher on eleven of twelve items assessing teaching ability than did those completing traditional four-year programs. Five-year program graduates also showed greater satisfaction with their teaching careers, had lower attrition from the profession, and indicated that they planned to remain in teaching longer than did those with a four-year teaching certificate.[1]

Recognizing the strength of the five-year model, universities such as Rutgers, the University of Virginia, William and Mary, West Virginia University, and Truman State now require a major in a discipline as part of teacher certification, believing, as West Virginia's Benedum Collaborative five-year Teacher Education program declares, "Traditional four-year teacher education programs across the country typically fail to address adequately the issues raised by current research on the knowledge base and professional preparation needed for effective teaching." A content area major is prescribed for each student planning to teach at the secondary level, and the program Web site includes a number of appropriate "double major" degree choices for those wishing to teach in the elementary grades.[2]

At Rutgers in New Jersey, teacher certification is added to a bachelor of arts degree in a discipline, or can be combined with an undergraduate major as part of a five-year masters in teaching. The university declares that "The two pillars of the Rutgers Camden Teacher Preparation Program are solid knowledge of a discipline and extensive practical experience. We produce students who know what they are teaching about and also know how to teach what they know."[3]

With data so strongly demonstrating that talented teachers can be recruited from among those who hold only a content-specific degree, who are then provided with training in pedagogy once employed, a starting point for the transition to a teaching force of content-specialists is to insist on a degree in a discipline as a condition of employment. Schools must at the same time begin the process of pressuring colleges of education and state certification agencies to include a bachelors of arts or science as a pre-certification requirement. As districts limit hiring only to those with a content-specific degree, universities and departments will get the message and requirements will change to meet hiring demands. But what of the thousands of teachers, many with great classroom skills, who lack an academic major and would become more effective in the classroom with a stronger content background?

Professional Development to Build Knowledge

Phase One of a robust district professional development program: turn each teacher into a content-specialist, beginning with middle school and high school faculty who do not have degrees in the disciplines they teach. Assisted by the evaluating administrator, teachers can chart an academic plan, develop a timeframe for completing it, and make progress toward completion of the plan a part of the annual evaluation process. Elementary faculty can be given greater latitude to select a discipline based on areas of special interest, but the purpose of achieving this objective is simple and straight forward—insuring that every teacher in the district has at least a bachelor's degree in an academic discipline within a specified number of years.

Although districts do receive funding each year for professional development, it will probably not be sufficient to support a transformational development program of this magnitude. Part of the greater community effort to increase salary support should include strengthening the budget for professional development for faculty through either committing an appropriate percentage of revenue to employee training or creating a foundation that solicits support from philanthropic individuals or organizations for this specific purpose. Athletic booster clubs manage to refurbish stadiums, outfit teams, and support expanded coaching staffs. If a public priority is to have children who can compete in the stadium of the global marketplace, the community must be equally willing to support the training program needed to develop world-class teachers.

Teach for America and the New Teacher Project both have demonstrated that even the brightest new college graduates with impressive credentials in an academic discipline benefit from additional training in instructional methods and classroom management. Their training programs have also shown that these skills can be effectively acquired "on the job," through a mandatory and rigorous new teacher institute. The first five weeks of a *Teach for America* recruit's career are spent teaching summer school in the summer before the first full-time term begins. Mornings and early afternoons are committed to the regular classroom under the supervision of a master teacher from the district and a *Teach for America* supervisor. During late afternoons, the rookies gather for their own learning in an institute, which they describe as "intense, challenging and rewarding," participating in short courses that provide further content knowledge and instruction in learning theory, classroom management, instructional planning, and literacy education. Two-thirds of the principals surveyed by *Teach for America* found this training to be more effective than that received by their other novice teachers, most of whom were entering from traditional teacher education programs.[4]

Preparing Teachers to Teach

Here again, regions within a state have an opportunity to bring districts together to form a New Teacher Summer Institute. All novice teachers, selected with primary emphasis on content preparation, instruct in summer schools under the supervision of the most successful master teachers the regional collaborative can find, while receiving afternoon instruction in the art and science of effective student engagement. A district or regional specialist trained in effective assessment and teaching strategies can take the place of the *Teach for America* supervisor and remain with the novices through the first several years, providing a period of consistent training and evaluation before the new employees leave probationary status and are extended two-year rolling contacts.

Continued status must mean continued training, focusing on currency in the teacher's discipline, improved pedagogy guided by an annual evaluation, and broadened understanding of the world in which students will live and work. Most current evaluation systems that choose not to take student achievement into account fail in two respects: they are unable to give teachers useful feedback on their own performance, and they provide no diagnostic tools through which a faculty member can determine personal learning needs. Data that show that an elementary teacher's students consistently do poorly in reading are as helpful to the teacher as to the students—indicating where targeted learning for the instructor can affect the learning of all. High school history faculty who find that their students, like my fifteen-year-old son, have little understanding of the political dynamics of Southeast Asia that led to the Vietnam War will know that both they and their students will benefit from a brushup on the global geopolitical realities of the 1960s and 1970s.

As technology in general, and particularly new instructional technology, evolves and becomes a ubiquitous part of our lives, ongoing teacher training also must provide teachers with familiarity and proficiency in its use, even with popular forms such as handheld devices and gaming. In early grades, students have acquired proficiency in these skills and expect teachers to be familiar and to use them effectively and with ease. When asked how he felt about teachers using technology in the classroom, one college student observed, "We're getting to a point where we're expecting the use of technology; it's becoming the standard," with another complaining, "but they're not always even able to run the computer or DVD player efficiently."[5]

James Paul Gee observes how effectively gaming technology promotes mastery in learning, with players forced to master a set of skills or complete complex tasks before moving on to the next level. They spend hours unraveling complicated puzzles, honing skills, and memorizing sequences, just to be able to progress through the game. He notes that "Kids today are seeing more power-performanced learning in their popular culture than they're seeing in their schools. When they go to high school and college, they're going to demand that we do at least as well. They are going to say, 'Why am I sitting here for this lecture (thirteen lectures, thirteen weeks for every subject matter I'm taking) in a room with the same people when this form of learning is so out of kilter with what I've seen in other spaces where it's worked much better?' "[6]

There is a profound irony, technology futurist Mark Milliron points out, in the fact that we now have a generation of learners who communicate, investigate, play, and socialize using handheld information managers, but our first instructions at the beginning of a class period are to either leave the devices outside of the room or make sure they are off.[7] Imagine how much more interesting, connected, and useful a class session might be if, instead of ignoring what has become the connective tool for the students' life outside of school, the teacher says "okay, turn on your handhelds. Today we are going to look at changes in demography around the world, and I want you to find the five largest cities in 1950, 1975, 2000, 2005, and projections for 2015. You have ten minutes to retrieve what information you can, then I'll put you in groups of three to match data, and reach a consensus. If you don't have a PDA, get to one of the laptops along the wall, and have at it." As games are developed that teach important lessons in the science, and as the computer lab down the hall becomes obsolete because each student has more capacity in her purse, teachers need to be ready to utilize these powerful tools. Gee adds:

> If you want somebody to do biology, and you want students to see how the biological words relate to actual experiences, games let you simulate those experiences as activities that people do. So you are actually getting ready for action. The best way of understanding a word is seeing how it applies in the world, how it applies when you have to do something. You have to use it in talk, or you have to use it in action. That's a fundamental principle of language acquisition. Games are good at that.[8]

We now also live in an age in which failure to use and understand technology cripples a teachers' ability to remain current. As research is increasingly published solely online, as newspapers discontinue their newsprint dailies and go entirely to web editions, and as Google rushes to convert the world of print to electronic format, our knowledge-driven society becomes a technology-driven society. The old traditionalist who refuses to take his World Civilizations class online to search databases for translations of the Rosetta Stone, choosing instead to send them a few at a time to thumb through Britannicas in the library's reference section, may himself be speaking a dead language to his students. Too many faculty still think of data storage and retrieval as being place-bound in volumes or in "learning center" rooms—when the vast majority of people don't look for information that way anymore.

Bringing the Outside In

Teach for America bases its teacher-training program on a series of leadership principles, believing that teaching is essentially an exercise in leadership. And a teacher exercising leadership expects to be at least as conversant as the student is in the latest learning modalities. Teaching as leadership also implies that the teacher is prepared to guide her charges into exploration of those areas beyond the classroom that will most dramatically affect the students' lives. Harvard's Business Professor Emeritus, John Kotter, outlines eight principles for all leaders in a change-driven environment— the first of which is "establish a sense of urgency." In more recent work, Kotter amplifies this principle by suggesting that an initial step to creating this sense is to "bring the outside in," openly examining what is influencing the institution and its future in the world beyond the organization. Describing the problem in business terms, Kotter writes:

> I can still remember a visit I made to a company ten years ago as if it happened last week. The firm had been exceptionally successful in the middle part of the twentieth century. But by the time of my

visit, market share had been steadily eroding for twenty years. The company was in a war, badly wounded, some would say bleeding to death. Yet when I opened the door to corporate headquarters, I entered a visual fantasy world.

Nowhere was there a single sign that the company was struggling, had been struggling for two decades, and was continuing to be beaten again and again in the marketplace. Nowhere was there a sign that the technology affecting its products was changing faster, offering it wonderful opportunities to leap ahead of competitors. The huge waiting room was pin-drop quiet and had the air of an antechamber outside the king's throne room.[9]

Kotter might just as easily have been describing some of our schools. I spend a good portion of every year talking to educators about the conditions outlined in the ETS *America's Perfect Storm* report, and the dire shortcomings of our workforce preparation discussed in *Tough Choices or Tough Times*. The intent in each case is to create a sense of urgency. One recent presentation was immediately followed by a large discussion session in which it was clear that a real sense of urgency *did* exist, and faculty anxiously discussed ways to convey this sense to their students. But often I am greeted with polite tolerance, accompanied by quick glances and winks between teachers that suggest some sympathy for the delusional alarmist who has been invited to fill the schedule on a mandatory "all staff" day. On occasion, someone is bold enough to tell me that he has heard it all before—that when he entered the teaching profession thirty years ago, the sky was falling, and apparently still is. I am not the first Chicken Little to come his way, and won't be the last. Others react much like some of the students in a Comparative Religion course I taught some years ago. After what I thought was a particularly enlightening discussion of the evolution of doctrine during the first four centuries of the Christian Era, several students would drop the class, telling me in a brief e-mail message that they simply "didn't want to think about it." But there are always several in the audience who come up following the presentation on *America's Perfect Storm* to tell me that they were in Shanghai or Bangalore the summer before and know exactly what I am talking about. Their comments remind me that, as we work to create a sense of urgency in our schools, there is a difference of significant consequence between those who have only "knowledge about" and those who have "experience with."

The year I mentioned earlier when my son Paul and I went to Thailand together with my friend Steve, while he spent the first week in Bangkok at the gem show, Paul and I did the touristy things in and around the capital city and in Chiang Mai to the north. We toured the Royal Palace, took a long-tailed speed boat into the famous floating market at Damnoen Saduak, rode elephants through the jungle, and rafted along a river to the site of an exotic Buddhist shrine cave. For Paul, Thailand was a giant Epcot Center where at night we didn't have to leave the park.

In the second week, we traveled to Mae Sot, a mid-sized community huddled along the Moei River that forms the border with Myanmar (Burma, during the colonial era) and one of the locations where Steve had been told gems were smuggled into Thailand. Much of the border is defined by sprawling makeshift camps of Karen tribal refugees, fleeing persecution by Myanmar's military Junta. As we walked through the camps, children with shaved, sore-infested heads seemed unperturbed by the flies that crawled around their eyes and noses as they squatted naked beside small charcoal cook fires, waiting for the handful of rice that would be their meal for the day. Family after family huddled side by side in makeshift shacks of tin or woven palm matting. Steve had been told that by searching these camps we would find places where the refugees spread contraband gems out on blankets to attract prospective buyers—a claim that turned out to be true only for very poor quality roughly cut stones. But the search took us through the heart of some of the poorest living conditions I had seen.

During our last day, we walked out onto a main street that ran along the river and were approached by a small group of lepers. The woman leading the tattered band appeared to be in her twenties, but was in such advanced stages of the disease that it had eaten away her fingers, lips, and nose. In one arm she clutched a tiny baby who lulled against her bandaged neck, its stomach distended by hunger. As the group approached, the woman held out a cup, its handle hooked over the stub of a thumb, and asked for alms. We fished into our pockets and gave her the small bills and change we were carrying, then moved on down the street. As we headed toward another cluster of corrugated metal sheds, I noticed that Paul had become uncharacteristically quiet, and when I glanced over at him, I noticed a tear welling in one eye.

"Are you okay," I asked?

He brushed away the tear and looked back up the dusty road at the retreating lepers. "I didn't know about this," he said. "...that people really have to live this way."

During Paul's fifteen years, was he taught about the squalid poverty that curses parts of our planet? Yes, he was. Did he learn about the scourges of disease? Yes, he did. He had "knowledge about" them, but it wasn't until he looked into the disfigured face of the leprous woman on the Moei River Road in western Thailand that he had "experience with" the grinding poverty and sickness that defines the lives of millions. Those few minutes changed his life—and also his sense of the urgency of doing something about it.

Learning Through Travel

I saw the same transformation as I led a group of teachers during a five-week Fulbright-Hays summer study experience in Syria and Turkey. As our guide, a Syrian Christian from the Golan Heights, told about his village being razed by the Israelis to establish a buffer zone along their northeast border, his own eyes misted, and he said, "To me it isn't a buffer zone. It is where my ancestors lived for many generations. And I can't go back there now—even to visit." In a Palestinian refugee camp a few miles away, an ancient woman who had been relocated to the camp in 1947 as a young girl spontaneously threw her arms around one of the young women in our group and thanked her for coming to visit. "We should be meeting together more often," she said in Arabic through our interpreter, flashing a toothless smile. In the heart of Damascus, we celebrated Shabbat with members of one of the oldest remaining Jewish communities in the world, while at the end of the alley leading to the unobtrusive synagogue, an armed Syrian policeman studied every movement in the street.

"Is he there to protect you, or to keep an eye on you?" I asked the old man who conducted the service. "Yes," he said.

The twelve members of our group returned to our classrooms different people, with a much more circumspect view of the complexities of a conflict that shapes our lives on a daily basis, and as much better teachers.

If we are to expect our own teachers to work with students as effectively as the two young Dutch women in the Cambodian market, we need to provide them with the same mind- and heart-expanding experiences. Beth Janssen, a high school art teacher in Wichita, Kansas, who was part of the summer Syria-Turkey experience, commented that nothing had influenced her teaching as significantly as had the five weeks in the Middle East. She observed:

> I not only see Art differently—I see the world differently. And I understand what students need to know about it differently. Even though we live in a global society with global ideas and communication, we still have *many* pockets of homogenous groups (i.e. suburbia and "white flight"). I still have students who only experience the outside world through the internet, but never think about how people all over the world are living lives that are so different, yet have so many life experiences that make us the same. It is this lesson I've learned from my travels. And it is this lesson alone, when learned by everyone, that will create a better, peaceful, global society.[10]

Brian Croone, a World History teacher in Stillwater, Minnesota, takes advantage of study abroad opportunities wherever he finds them. In addition to being part of the group to the Near East, his travels have taken him to South Africa and to the Balkans. "History is a living thing in the sense that it influences everything that is happening in the present," he says. "To understand that impact, you need to experience it where it is happening." Like the young Dutch teachers, he sees travel as essential to helping his own students connect with the world. He observed:

> Travel abroad opportunities really keep teaching exciting and new
> for me. In addition to learning about other cultures and seeing things
> I might not otherwise see, it's amazing how many times you can
> relate what you are doing in the classroom to a story or an incident
> you learned while traveling. It is these little stories that get students
> excited about a subject and about their own possibilities for overseas
> travel—making them want to learn more about other cultures and
> peoples. The travel I have done helps students see beyond the myths
> and stereotypes we have about other countries.[11]

Brian claims that it is not difficult to find opportunities to travel abroad
as a teacher and to have the expenses covered. He learned about some
of the programs through presentations at the National Social Sciences
Convention, and looks for others on the Internet. Once fellow teachers or former study abroad colleagues learned of his interest in further
travel, they sent notices or called about opportunities. "I heard about
your program in Syria and Turkey from a teacher that I was with on a
trip to China," he said. "A friend told him, and he in turn told me. So
once you have done any kind of program you get connected with people
that have done similar trips." Coincidentally, on the day I was working on
this section of the book, I received an e-mail from Toyota's Institute of
International Education, announcing "fully-funded, international, professional development opportunities for U.S. educators....the program
aims to advance environmental stewardship and global connectedness in
U.S. schools and communities through creative, interdisciplinary, and
solution-based teaching methods."[12] What more could a district ask than
a rich, content-centered professional development opportunity for faculty
at no cost to the school!

Phase Two of each district's professional development program should
be to assist each teacher with a meaningful international experience. Not
a week in a self-contained, full-services resort in Cancun where there
is always a buffet beside the pool and a waiter checking every few minutes to see if a margarita needs to be freshened, but a couple of weeks
in an eco-camp in Costa Rica or studying the influences of Islam in
medieval Spain. The Fulbright-Hays program takes hundreds of teachers
abroad each summer and Brian's Balkans expedition was sponsored by
The Center for Middle Eastern Studies at the University of Arizona. This
"awareness" goal can be pursued with relatively little district expense
since, like the Toyota experience, the Fulbright-Hays programs cover
all costs and are usually designed to give teachers from specific disciplinary areas both authentic international experience and rich content-
related exposures. But too many of these participants are "repeaters,"
who acknowledge that their K-12 colleagues—even when all expenses
are paid—choose not to leave the country unless as part of a carefully

packaged, full-comfort tour. To change this over time, in addition to "bachelors in a discipline required," position announcements for new teachers should include an indication that "International experience is preferred." Our second message to colleges of education needs to be "get your students out of the country for a meaningful study abroad experience, because we want them to know the world about which they will be teaching, and to exhibit a sense of urgency."

In a short video that supports one of his books on Amazon, John Kotter declares:

> Urgency is fundamentally an attitude; a way of thinking, a way of feeling, and a way of behaving. It's a sense that the world out there has enormous opportunities and hazards. It's this gut level determination that we've got to deal with that. I've got to do something today, each and every day, get up with this "I'm going to do something that's important." .... It's this determination not just to have a meeting today, it's to have a meeting that accomplishes something...that moves us forward on important stuff.[13]

For teachers, this means entering the classroom each morning committed to doing something important with that day—to knowing that, when the last student is gone and the final duty completed, something has been accomplished that moved students forward on "important stuff." Paraphrasing the objectives of teacher education at Rutgers, each district should be able to say about its professional development program, "The three pillars of our Teacher Preparation Program are solid knowledge of a discipline, extensive practical experience, and a sense of urgency created by firsthand experience with the world in which students will work and live. We produce students who know what they are teaching about and also know how to teach what they know."

Include School Leadership

The often unsung hero and critical ingredient in the formula to select, develop, and inspire great teachers is the building principal who, recent research indicates, has about a fifty-fifty chance of not lasting in the position for more than five years.[14] As schools improve, disciplinary challenges decrease, teacher quality rises, and boards and superintendents heap on support, turnover should begin to decline. But professional growth opportunities mustn't end when a talented teacher moves into administration. The success with which a school can develop a highly effective teaching corps, assess and improve performance, and create a growth-oriented development program will depend on keeping principals in positions long enough to affect and sustain change. No one would be more acutely aware of the sense of urgency, or better equipped to translate it into action for those doing the work in the classroom, than our principals. They must be extended the same opportunities for growth and the same encouragement to get out and experience the world.

The Talmud admonishes that "The day is short, and the task is great, and the workmen are sluggish, and the reward is much, and the Master of the house is urgent."[15] For us, most of the workers are *not* sluggish, and are anxious to work as effectively as possible if equipped with the skills and understanding needed to do the job. But the day is short and the labor long. The reward is beyond measure, and the Master—our social, cultural, and economic future—is indeed urgent.

CHAPTER TWELVE

Reinterpreting "Least Restrictive Environment"

"It used to be believed that the parent had unlimited claims on the child and rights over him. In a truer view of the matter, we are coming to see that the rights are on the side of the child and the duties on the side of the parent."

—William G. Sumner

In a small rural school in the upper Midwest, eleven-year-old Courtney has been placed in a regular sixth grade classroom, despite the serious challenges created by severe emotional and behavioral disorders and a history of violent outbursts at school. Although assigned a full-time aide, when Courtney has an "outburst" her sixth grade teacher has been instructed to remove all the students from the classroom to wait in the hall until Courtney can be calmed to the point that her teacher and aide feel she is no longer potentially harmful to herself or her classmates.

In a neighboring district, ten-year-old Jason also has a full-time aide and has been diagnosed with acute Attention Deficit Disorder that causes frequent yelling outbursts and spontaneous running around the classroom. Because of these outbursts, Jason's Individual Education Plan (IEP) has been changed to limit his time in specialized classes such as music and art to only ten minutes at a stretch, cycling him in and out of the classes for brief intervals that accommodate his attention span.

Least Restrictive Environment

Something akin to these examples plays out in school districts across the United States on a daily basis as teachers and school administrators struggle to comply with requirements of a law they often refer to as "Ninety-four one forty-two," specifically Public Law 94-142 signed by President Gerald Ford in November of 1975. Initially titled the Education for All Handicapped Children Act, the law has since been renamed the Individuals with Disabilities Education Act (IDEA) and builds upon provisions of Section 504 of the Rehabilitation Act of 1973, outlining specific requirements related to education. Section 504 states:

> No otherwise qualified individual with a disability in the United States...shall, solely by reason of her or his disability, be excluded from participating in, be denied the benefits of, or be subjected to discrimination under any program or activity receiving Federal financial assistance.[1]

Since public schools universally accept support from the federal government, all are subject to the provisions of Section 504 and of IDEA. The laws mandate that all school districts must identify disabled children living within their jurisdictions and must provide "free and appropriate education" regardless of the nature or severity of the handicap. To this provision, the IDEA legislation adds the requirement that special-needs students' education must be provided within the "least restrictive environment" and, in combination, the statutes make districts responsible for insuring that resources needed to meet these expectations are available, and impose sanctions (withholding of federal funding) on districts found to be out of compliance.

These dramatic protections of the rights of children with disabilities were in large part a response to a national disgrace—the revelation that, as late as the 1970s, of eight million children in the United States with disabilities, half were receiving no educational services.[2] But while requiring that students be "mainstreamed" into regular classrooms where practical, and educated in a "least restrictive environment," the legislation provided little guidance as to what these terms meant, leaving specific interpretation to the courts.

In a series of cases beginning with RONCKER v. WALTER in 1983, courts undertook the process of bringing greater definition to the IDEA requirement for a "least restrictive environment," or LRE. In this case the Court applied two tests—"(1) Can the educational services that make the segregated setting superior be feasibly provided in a nonsegregated setting? (If so, the segregated placement is inappropriate.) (2) Is the student being mainstreamed to the maximum extent appropriate?"[3] These tests provided general guidance to court decisions over the next half-decade, leading to a decision in the case of a kindergarten student who was assessed with an IQ of 40, that a "continuum of placement options" must be considered when determining LRE, beginning with mainstreamed placement in a regular classroom. Unless some compelling justification could be provided for segregating the student, she must be placed in a general education setting.[4]

Nearly a decade after passage of Public Law 94-142, the statute received perhaps its broadest interpretation by the courts when in the 1993 case of Rafael Oberti, an autistic student whose behavior had been judged disruptive to the rest of his class, the court ruled that "inclusion is a 'right,' not a privilege for a select few. Success in special schools and special classes does not lead to successful functioning in integrated society, which is clearly one of the goals of the IDEA."[5] Advocates for the disabled heralded the concept of full "inclusion," meaning that children with disabilities should be fully integrated into all aspects of the academic and nonacademic school day, and the term has become part of the common lexicon about how students with disabilities should be accommodated.

A year later, SACRAMENTO UNIFIED SCHOOL DISTRICT v. RACHEL H, a case involving a cognitively disabled girl named Rachel Holland, began to moderate the "inclusion as right" provision of OBERTI, outlining what is referred to as the Holland Test. This guideline created expectation that schools should look at three considerations when determining least restrictive environment: the educational benefits to the student of being included in the general education classroom vs. the benefits of a special classroom; the nonacademic benefits of interaction with nondisabled students; and the effect of the student's presence on the teacher and on other students.[6] In 1995 in an Arizona case, POOLAW v. PARKER UNIFIED SCHOOL DISTRICT, the 9th Circuit Court cited Holland in strengthened application of the test by finding that the benefits of inclusion had been adequately tested with several unsuccessful years for Poolaw in a regular classroom setting. The Court determined that the benefits of inclusion had been minimal, and that the student's needs could best be met by the residential placement recommended by the district.[7]

Best Interest of All Children

In a number of cases since the Holland Test was established, courts have ruled that "Disruptive behavior that significantly impairs the education of others strongly suggests a mainstream placement is no longer appropriate."[8] In each of these cases, however, the court was responding to a challenge by parents of the disabled student, not to concerns expressed by other parents that their children were failing to receive appropriate education because of constant disruption in the classroom. Yet a special education teacher whose own children were in just such a general education setting commented:

> I believe it is the single biggest factor in the decline of standards and achievement in public US education. I went to read a book in my son's kindergarten class, and there was a child in there who was so annoying during a 20-minute read that I wanted to canonize the teacher when I left. I would only describe him as seriously ADHD, but the inclusion of others with much more serious disabilities occurs all the time... There are benefits to students with and without disabilities when inclusion occurs. But, LRE is greatly abused, in my opinion, and all of the kids in the middle get sacrificed. What about THEIR educational rights?[9]

To get some unscientific measure about how common disruptive behavior created by LRE placement is on general classrooms, I spoke with two of my wife's former colleagues from her days as an elementary classroom teacher. Each now teaches in a separate district. I'll call the teachers Melinda and Janice, and one works primarily with special-needs students and the other is a general education classroom teacher. Both agreed that inclusion is definitely the right choice for most students with disabilities and that nondisabled children benefit from being in class with special-needs classmates. Some, particularly those with normal cognitive function but with various learning disabilities, physical limitations such as orthopedic handicaps, sight or hearing impairments, are among their better students. Data support that these students now complete high school at rates that are as high or higher than their nondisabled peers.[10] Even students with mild mental or emotional disorders can be managed in the general education classroom without negatively affecting the learning environment for other students. But each of these teachers also agreed that there are cases where "inclusion" is neither appropriate for the disabled student nor conducive to a healthy learning situation for others in the class.

The special-needs teacher, Melinda, noted that in a perfect situation every classroom teacher would be trained to work with students with disabilities and would be supported by both a special education teacher in the school and an appropriate ratio of aides in the classroom assigned to help students with special needs. But she acknowledged that it rarely happens. As a result, children with special needs who are placed in mainstreamed classes do not receive the support they need, and teachers in most general education classrooms are not equipped to manage their education. She believed very strongly that the majority of students with special needs needn't distract from the learning provided to nondisabled students, if the disabled students were adequately supported, but acknowledged that disruptive students do get placed in classroom situations where they negatively affect the learning of others. The law, she said, clearly allows schools to place these students in a separate situation for whatever parts of the school day are disrupted by the student's behavior.

Asked why disruptive students still remain in mainstreamed classrooms, Melinda said:

> I think there are generally two reasons. Parents and the advocacy groups can be very forceful—very intimidating. Sometimes administrators are unwilling to challenge them on the issue. But I think most of the time it is financial. If a school doesn't have the student in a regular classroom situation, they have to pay to place them somewhere else. We have just such a situation in my district. There is a very good center for special-needs children that is cooperatively supported by a number of school districts in the region. But the superintendent here doesn't want to pay the costs of sending the child there when budgets are so tight. His (the child's) parents haven't insisted,

so he stays in the general education classroom and it isn't good for any of the children.[11]

Janice, the general education classroom teacher, is not convinced that a student has to be disruptive to interfere with the learning of others in the class. "Some students with mental and emotional challenges that are pretty restricting don't have aides assigned to them," she said. "When you are teaching in a class with 28 to 30 students and several of them are not able to keep up with the pace of the class, the class slows down. I spend time reviewing, repeating, and I don't move on to new material as soon as I would otherwise—just making sure these students aren't too lost." She agreed that if adequate aides were available who could work with the students as she teaches, she could probably maintain the pace she would prefer. "But everyone needs to realize that these students won't achieve at the same level as others in the class—even with special assistance, and sometimes both parents and school administrators just don't want to admit that."[12]

As pressures continue from advocates for inclusion, a number of parents of disabled students are beginning to push back—arguing that their children are better served in separate classrooms, surrounded by other students with similar challenges and assisted by teachers and aides with special training with the disabled. In New Jersey, which has a long history of special schools for students with disabilities, State Legislator Steven Sweeney, whose daughter has Down syndrome and has suffered from the isolation and frustrations she experienced in an inclusive setting, is adamant that children and parents should have other options. "Just to put my child in a building to make people feel better because it's inclusion is outrageous," Sweeny said in an interview with the *Wall Street Journal*.[13] In the same article, Mary Kaplowitz, a Pennsylvania special education teacher who was once a strong advocate for inclusion, noted how important she now sees it for her son, Zachary, who is autistic and has mild cognitive dysfunction, to have classes in his local elementary school that are dedicated to students with similar disabilities. She found that Zachary was shunned and teased by classmates in general education classes and felt much more at home in a separate classroom.[14]

Broadening "Least Restrictive"

Without being insensitive to the needs of children with disabilities, parents with children in classrooms that are regularly disrupted by violent or uncontrollable behavior must also exercise their rights and those of their children to learn in a "least restrictive environment"—the one protected by the provisions of the court's Holland Test. Janice believes parents often do not successfully exercise this right for one of several reasons: they are not aware that their child's classroom is constantly being disrupted, they are afraid of the very vocal and aggressive inclusion advocacy groups, or worry about their own child being stigmatized by their request that a disabled child be placed in a separate setting. They also find, when they ask that school administrators move a continuously disruptive child, that school leaders are worried a lawsuit or a forced alternative placement will be an unmanageable expense for the school.

I am fully aware of the sensitivity of this issue and of our need to protect the rights of those who are least able to protect themselves. But if we are to create the best learning environment possible for *all* children, parents and administrators must be bold enough to insist that the provisions of the Holland test be applied in cases in which there is apparent disruption to learning. Though the wishes of parents of the disabled child must certainly be taken into account, the primary concern must be for the children—*all* children. We are a society that goes to great lengths to protect the rights of minorities, as we should. But we can certainly provide a free and appropriate education in the least restrictive "manageable" environment without limiting the learning opportunities for the vast majority of children.

A Role for Government

Here, without question, is an important role for state and federal government in education reform. If federal law mandates free, appropriate education in a least restrictive environment, this cannot be an unfunded mandate—a requirement of the law that is not accompanied by the resources needed to meet the expectations of the law. When it is determined that a special-needs child can function effectively in an inclusive environment with special training for the teacher, support of a colleague with extensive training in special education, and an adequate number of classroom aides to assist students with "keeping up," parents and schools must pressure government to provide these resources. When a child requires special classroom placement because behavior limits the ability of other students to function in their own "least restrictive environment," schools must be provided with the resources needed to cover the costs of this placement without having to pull funding from other critical instructional programs.

As communities undertake the challenges of creating internationally competitive schools, part of the review process should be an examination of appropriate placement of students with special needs, insuring that when possible they are given an opportunity to mainstream with non-disabled students. But when the requirement to care for student behavior limits the ability of the teacher or other students to appropriately teach and learn, other placement provisions should be made—and should be made affordable. If resources are not available to make these provisions possible, it is time for the district and community to mount a campaign similar to the one outlined in chapter 10 to pressure the government to support mandates with the resources needed to carry them out.

CHAPTER THIRTEEN

Providing Choice in the Public Arena

"The strongest principle of growth lies in human choice."
—George Eliot

When Ken Owen was notified that his North St. Francis County R-1 School District had been selected as one of five in Missouri to initiate its new A+ Schools program, he knew exactly where he wanted to begin. He wanted parents and students to have greater choice. The new state initiative was designed to encourage schools to get away from the traditional "generic" curriculum by creating options that improved college readiness or strengthened preparation for careers. Students who met the criteria outlined in the legislation—a minimum grade point, sixty hours of community service each year, 95 percent attendance, and completion of an adapted A+ curriculum—would be eligible for two years of tuition and fees support at one of the state's community or technical colleges. The regional UniTec Career Education Center sat right beside the high school in Owen's district, and he saw A+ as an opportunity to bolster its reputation in career education, while giving those who lacked the interest or ability to pursue a baccalaureate degree a viable and acceptable choice.

Owen knew that whichever direction students chose, they would need strong skills in math and reading, and part of the state grant that accompanied the A+ award went immediately to bolstering instruction in both areas, beginning at the elementary level. As curriculum was strengthened, the district placed emphasis on the importance of "application," with students having the option during the high school years to pursue a more career-focused curriculum through the UniTec Career Center. Short, professionally designed videos explained employment opportunities available to those selecting various career pathways with a message that "a four-year college degree isn't the right choice for everyone, but you will need some training beyond high school whatever path you take. If you are interested in a technical career, we have options that will give you practical experience as you take your classes, getting you ready for further career training." The district brought parents and guardians together to view and discuss these videos, demonstrating to parents and their children that the district had a program that was right for every student and that school could have value and meaning, whatever the student's goals. Summer school was redesigned to be much more than "catch-up" instruction for students who were behind. Technical "activity" courses involved students in inventing, redesigning, and exploring by utilizing concepts they had learned the year before, and summer school enrollment rocketed from 240 to 500.

"Some teachers didn't like the emphasis on applied learning and didn't want to have to create activities that engaged students in application of what they were learning," Owen said, "and we helped bring them on board or find other places to work. In those first years, we probably had to replace or retrain a dozen teachers who initially didn't want to deal with the change." But the program worked, and by the time it was fully implemented, dropout rates were cut in half, persistence to graduation rose by 9 percent, and the percentage of students completing advanced coursework more than doubled.[1] Disciplinary problems in the system as a whole declined by 67 percent.

"Students loved it, and when students are happy, parents are happy," Owen said. "Everyone was doing better academically, parental support went way up, and all it took was an acknowledgement that we can't have a cookie-cutter curriculum, and have to find something that excites every student about what he or she can accomplish."[2]

Even within the charter system, school administrators have recognized the value of extending personal choice. Sequoia Charter School in Phoenix initially created two options for its students: a "back to basics" curriculum and a more experimental projects-based model. Over time, these options expanded to five, recreating a series of mini-magnet schools concentrating on math and science, the arts and humanities, or the social and behavioral sciences within a single school. Each track is served by its own core of teachers and works to maintain the same standards of rigor and achievement, but with parents and students choosing the concentration that best matches the interests of the student.[3]

Creating Choice in the Public Setting

The second of the KIPP charter schools' five pillars for success is "Choice and Commitment," and states: "Students, their parents, and the faculty of each KIPP school *choose* to participate in the program. No one is assigned or forced to attend a KIPP school. Everyone must make and uphold a commitment to the school and to each other to put in the time and effort required to achieve success."[4] Therein lies the rub for school districts as a whole, in that most students do *not* choose to attend a particular school district, but live within it by circumstance. The district cannot simply say, "follow our rules or go live somewhere else." But the district can say, "We have developed curricular paths and to participate in paths A, B and C, here are the commitments you must make. If you are not willing to make those commitments, Path D is available to you." In many districts, charter schools have essentially created these paths, with Path D being non-charter participation. But the same choices can be offered within districts without formal chartering, by designating buildings or class sections as "Challenge" opportunities (or by some other name), with an expectation that those attending Challenge units agree to comply with the same standards and expectations that define the best charter programs: longer days and school years, commitment and hard work, written expressions of parental support, high standards of discipline, and regular assessment of progress. Within the Challenge options, curricular paths can afford the array of choices created by Ken Owen's North St. Francis County district or the Sequoia Charter school, with expectations remaining the same within each pathway choice. When students reach the secondary school level and cannot be divided by building or section, the common "alternative school" model provides an option for those who choose not to meet the expectations of the Challenge curriculum.

For some, this will smack of "tracking"—of imposing decisions about students' educational futures based on some established measure. And, essentially, it is. It is tracking based *not on ability* but on commitment. No student willing to attend regularly, adhere to discipline and effort expectations, and do his or her best work should be denied access to a Challenge program. Unless parents or guardians indicate they do not want their children to be part of the Challenge option, students should be allowed to make the choice themselves, even when those responsible for their care refuse to commit to the support statement requested by the school. Students and families who have elected not to participate should have the opportunity to opt back into the Challenge programs with an understanding that if the child has fallen behind his peers, extra work will be required to catch up.

In 1989, New Zealand, one of the countries where students routinely outperform their American peers, took the dramatic step of doing essentially what this book suggests—extending charter-like privileges to all of its public schools within the framework of national curriculum standards. Schools in New Zealand were granted much greater autonomy and parents were allowed to choose their children's school, with specific interest in encouraging schools to tailor curriculum to the needs of individual students. Judith Aitken of the country's Education Review Office noted that "Very few public education systems anywhere are highly regarded for their ability to think in terms of individual students, and supply individually-geared programs," and few countries are willing to address the national changes sought to address these concerns.[5] In the United States, we have been largely unsuccessful at extending both options and choice.

Developing a "Choice" Model

When in the middle of the last century economist Milton Freedman suggested the creation of school vouchers that would allow families to commit their education dollars wherever they felt their child would receive the best education, he believed that competition would eliminate weak schools over time, and would strengthen mediocre ones. In his model, choice should not be limited to public schools, but should be extended among all districts, public and private, and quality should drive the market. With a few exceptions, vouchers have gained limited traction over the past fifty years, largely because of concerns from those on the political left that it will result in a two-tiered school system in which families who value education and those who can supplement their vouchers to afford higher-cost private schools will flee weak public institutions. Weakened public schools will not collapse or disappear, but will remain detention centers for the unmotivated, under-supported, and undisciplined. Some on the right fear that vouchers may lead to a loss of institutional autonomy as government inserts itself into curriculum choice, disciplinary policy, and governance structure. The alternative solution to creating choice that drives quality standards while avoiding the divisions created by cost and family support is to create those options *within* public districts and buildings, rather than across districts and across public and private sectors.

Hardwood Unified School District in an upper New England state is my creation to illustrate how choice options might be developed and integrated within a public district, leading eventually to a full Challenge model. Hardwood is a midsized district with a single high school, middle school, and five elementary buildings. Oak and Ash Elementaries begin the change process by becoming Challenge schools, one offering an immersion Spanish section at the second grade level, and one introducing Chinese. Maple, Yew, and Beech maintain traditional schedules and serve as the alternative choice. (This ratio can differ, of course, based on how successfully the board, the CORE Team, and its support groups persuade families to select the Challenge option.) Challenge schools will move to a 200-day calendar, lengthen the day to create longer time-on-task in key academic areas, require uniforms, introduce a mandatory two-to-five-week summer term, and solicit written statements of parental support and participation for every attendee. Teachers with the strongest academic credentials and the best performance records are moved to Oak and Ash, and as new teachers are hired, only those with a degree in a discipline are considered for these schools. If the state allows alternative teacher certification or if reform activists have been successful in getting chartering privileges extended through legislation, every new teacher participates in the district's Summer Teacher Institute, where the most talented recruits are identified and assigned to Oak and Ash. Salaries for teachers working in these extended program schools are adjusted accordingly, with enough bonus incentive to make these the most desirable teaching positions in the district. In essence, two charter choice options are created, with families making the choices.

As students from Oak and Ash move into Teak Middle School, the district adds a Challenge option, with sections requiring the same extended schedule, parental commitment, and academic expectations that Challenge students bring with them from the lower grades. Students entering from Maple, Hickory, and Yew can choose to enroll in the Challenge program at the Middle School, but may need a summer transition experience (part of the Summer Teachers Institute) to insure they are at grade level with their Oak and Ash peers. Special weekend excursions and summer opportunities become part of Teak's Challenge curriculum, as do advanced classes in Spanish and Chinese, adding to the sense that "this is where families want their children to be."

At Ebony High and at Mahogany Career Center, the Challenge option continues with teacher pay differentials, stricter teacher preparation requirements, separate class sections, and distinguishing uniforms for those in the Challenge program. Just as with a district having a number of excellent charter schools, Hardwood district has created a two-tiered system—but one based entirely upon choice, effort, and commitment rather than ability. The goal over time is to bring Maple and Yew into the Challenge fold with a corps of equally prepared and dedicated teachers, eventually reducing Hickory to a small alternative program for those with chronic behavioral problems or who simply refuse to meet standards of attendance or work commitment. As more Challenge sections are needed at Teak, Mahogany, and Ebony, they also transition fully into Challenge schools, with a small alternative program for those who will not or cannot meet the behavioral or effort expectations of the Challenge programs. By this time, those teaching in the alternative program will also be among the district's star performers: teachers who have demonstrated special talent for working with difficult and under-motivated populations—always with a goal of encouraging students over into the Challenge track.

Notice that the operative word here is *commitment*, not *ability*. As long as a student applies himself and turns in his best work, he has a place in the Challenge program, though he may be getting substandard grades. Parents constantly will need to be reminded that being in the Challenge program is not a guarantee of academic success. It is simply a guarantee that the district is providing the best learning environment it can create for the child, and that the student will achieve a higher level than she would have under the old traditional program, given her talents and effort.

This model clearly is designed to separate and create distinctions in opportunity. But reform will not occur unless distinctions between the new and old are clear and visible to all. If I cannot see that my child is missing something important and worthwhile—something that I can provide her by simply saying, "Yes, we will do this," I will not change. But I *will* change if it becomes visibly apparent to me that another accessible choice exists, and that my child is losing advantage by not being part of it. The clamor to move students into the best charter schools demonstrates this principle every year, and we are robbing many of our children of the greatest future opportunities by failing to provide them with choice.

Reorganizing a district or buildings in this way may appear to be an administrative nightmare, as teams struggle to determine how to separate Challenge sections from traditional classes, and to move students around the district to different elementary buildings based on parental choice or language interest. But districts such as Los Angeles, one of the largest in the country, are experimenting with intra-district choice options, and those with charter schools have managed similar reorganization, so it can be done. One of the greatest enemies of reform is our willingness to allow existing processes to impede change decisions—yielding too easily to the "we can't do that" voices whose justifications are usually no more than masks for not wanting to do the creative thinking or put forth the effort. The charter movement demonstrates that districts *can* be divided by family choice, and that students can be moved to accommodate these decisions. The only real limitation elsewhere is the absence of imagination and will.

CHAPTER FOURTEEN

Legislating for Change

"The dogmas of the quiet past are inadequate to the stormy present. The occasion is piled high with difficulty, and we must rise with the occasion. As our case is new, so we must think anew and act anew."

—Abraham Lincoln

From our discussion up to this point, it should be clear that as a community begins to create a culture of education and assume responsibility for strengthening its public schools, there is a great deal that can be done without charter school privileges. Curriculum can be changed, the senior year can be strengthened, calendars and school days can be adjusted, and students can be held to higher standards of behavior and academic achievement. But each district will reach a point where, without the freedoms chartering legislation can provide, some of the more challenging problems involved in reform will be difficult, if not impossible, to address. Central to these issues are those involving teacher and administrative selection, certification, evaluation, and termination.

Tracing Reform Rhetoric

A fascinating feature of several of the online search engines such as Google is that they have created timelines for various historic trends by charting the amount of print coverage a topic received in any particular year. These timelines are often depicted as bar graphs spread over one or two centuries, dating back as far as newsprint records are available. By selecting a decade or other designated block of time, a mouse click brings up a listing of both the frequency and types of articles written about the topic during that period.

By tracing the history of education reform in America through this process, a timeline comes up that dates back to 1820, and is relatively uneventful for the first hundred and sixty years. There is a modest bump in the 1960s and into the mid-1970s. Then in 1983 the chart spikes, catches

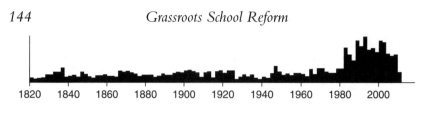

Figure 14.1   Print coverage of education reform in America

its breath for a year, and from then on shows steady activity through 2000, followed by a gradual decline (see figure 14.1). (A link to this table is provided in the endnotes.)[1]

So, what happened in 1983 that caused the profile to jump? The *Nation at Risk* report was published. But the problem with our timeline after that year is that it is not a graph of the magnitude of reform, but of the amount of public discussion taking place *about* reform. As the data reviewed in earlier chapters illustrated, not much improved in American education over that quarter century, and I would attribute much of this intransigence to a general unwillingness of communities to accept responsibility for their own performance. They allowed the power of local, state, and national organizations representing teachers—groups that have actively resisted major efforts to base faculty selection, assessment, retention, and dismissal on student performance—to determine policy and practice. These organizations have been strident voices in support of tenure for K–12 teachers, which is a system of protections that serves little purpose beyond making it difficult to dismiss teachers who are not producing results. In order to most effectively reform education in America, districts must have authority that goes beyond adjustments in calendar and curriculum, and must have authority to hire, fire, assess, and evaluate based on how successfully faculty produce results in student achievement. This will not happen without new state policies that extend to each school district the authority equivalent to that provided by the better charter laws.

A Case Study in Legislative Change

There are numerous references throughout this book to community colleges—partly because they are one of America's truly great educational experiments and achievements, and partly because they are a fascinating case study in how a movement can gain momentum and sweep across the nation. The first public community college still in operation appeared on the scene in Joliet, Illinois, in 1901 and was followed a decade later by legislation in California that allowed high schools to add two years to the traditional twelve to serve as the freshman and sophomore years of a baccalaureate degree. By the end of the century, approximately twelve-hundred two-year public colleges were strategically located across the United States, positioned so that 90 percent of the population lived within a comfortable commute. The 1960s was a decade of spectacular growth in the community college sector, with a new institution springing up on an average of one per week during that entire decade. Between 1963, when the Center for Education Statistics began keeping data on community college enrollment, and 1970, student numbers in this sector grew from 850,000 to 2.3 million, and now top 6.7 million. America had taken hold of a reform idea that worked, and wasn't letting go.

What was it about the idea of the community college that had such appeal, and that allowed communities to successfully petition state legislatures to change both statute and departmental policies in dramatic ways, and in a relatively short period of time? As I have spoken with the early founders of "the movement" who were pushing for legislative action during the community college boom days of the Sixties, most say that one of the most critical factors was that "it was an idea whose time had come." Following the end of World War II, with thousands of GIs returning home and a new GI Bill offering benefits that supported additional education, there was a need for both greater access and capacity in American higher education, and greater variety in training opportunities. A commission assembled by the Truman Administration called for creation of a network of "community colleges" to fill this need, and the nation began to respond. In the late 1950s, with the launch of Sputnik, the beginning of the Space Race with the Soviet Union, and a new generation of Korean War veterans entering college and the work force, education in general gained a greater sense of importance for all Americans, and no parent wanted his or her child to miss out on the opportunities afforded by a college degree.

Jim Tatum is a graduate of West Point and veteran of the Korean War who, after being wounded in combat, returned to manage a business in southwest Missouri. In the early 1960s he became part of a small group of community leaders from his home area and from St. Louis who launched the effort to create community college legislation for the state. I met with Jim at his home overlooking a beautiful valley in the Missouri Ozarks to learn about the process of successfully championing legislative reform.

"First," he said, "you have to help people see that this movement is in their own best interest. It can't be your idea. It has to be theirs." But in the early 1960s, the idea of creating a new college in the region faced one significant hurdle. It required an increase in local taxes—a tough sell in the rural Ozarks.

"I started," Jim said, "by talking to everyone I could about the future of our area...about what they would like for their children. When they would say 'we need better jobs or more opportunities,' I asked, 'what do you think we need in our area to make that happen?' These are bright people and pretty soon someone suggested, 'it would be nice if we had one of those junior colleges.' From then on as I spoke to people, I could say 'Frank and Caroline mentioned that they think we should do what we can to establish a junior college. What do you think of that idea?' Over time, people were pushing me to support *their* ideas about starting a junior college."

In St. Louis the same discussions were being held and groups in each region found a key state representative willing to draft legislation that would allow contiguous school districts to hold elections to create junior college districts. The legislation had two characteristics that were especially appealing to state politicians. The first was that it was just "enabling." It didn't actually create these new districts but provided mechanisms through which local citizens could do so. At the same time, it described what these institutions needed to look like, once created. It didn't force new colleges on the public, but outlined what one must include in terms of governance, general program approval, financial support, and so on when one was created. In the district proposed by Jim Tatum and his colleagues, the vote carried by 70 percent.

Giving Birth to the Idea

As CORE Teams begin their work in reform-minded communities, the first task is to re-create that spike in print coverage and public attention that occurred in 1983. As reports such as ETS' *America's Perfect Storm* and *Tough Choices or Tough Times* are published, a CORE Team member can be assigned to write an Op-Ed piece for the local paper highlighting the findings. School boards should hold public forums to discuss the future of education and of opportunities for the community's children, reinforcing the message that, as other countries become better educated and have lower wage scales than the United States, it is inevitable that work will go elsewhere. And though we can't change the rate of improved education in other countries, we can change our own. Then a series of one-on-one conversations must begin throughout the town or city, similar to those initiated by Jim and his colleagues. "What do you think we need to do to remain competitive and to bring new economic vitality to our community?" As people suggest that one of the critical needs is to strengthen public schools, and as key members of the community visit successful charter examples, the charter idea will inevitably surface. "Why shouldn't our local school have the same freedoms and privileges extended to these charter schools," the CORE Team member might ask? "Why, I think they should!," Frank and Caroline will say, and suddenly the idea of extending charter opportunities to the full district becomes the community's idea. A state representative friendly to the proposal can get the legislation drafted, and the ball is set rolling.

Obviously, it isn't that simple. The community conversation will need to be well planned, carefully managed, and continuous. Though any legislator can draft a bill, the process of moving one successfully through a state legislature is difficult at best, and often requires several years of persistent lobbying and public pressure. As one of the Fundamental Laws of the Political Universe noted earlier, "It is much easier to kill a bill than to pass one." Bills are assigned to a committee by powerful House Speakers who have a number of ways to keep them from ever seeing the light of day. They can assign them to an unfriendly committee, asking the committee chairperson (who the speaker probably appointed) to keep them "off the calendar," meaning that bills never come up for formal debate. Or the Speaker can let a bill move through the committee process, but keep it from coming up on the general legislative calendar. Key to the bill's success is creating a proposal that doesn't immediately generate widespread opposition, then cultivating the right support in the right places.

In states with no charter school laws, the initial step should be to draft a model charter schools bill, and include in the legislation that any school district can, by vote of its school board, elect to operate under the charter provisions. The bill, and any legislation drafted to move this reform initiative forward, must also include requirements that the district document student performance, and within a reasonable period of time demonstrate improvement in areas of critical learning if the charter is to be retained. Rather than attempt to introduce a new assessment system to the state, those drafting the legislation should use existing testing requirements, leaving it to the schools to rise above the limitations that a testing structure might place on curriculum and breadth of content. To be successful and to avoid the problems and criticisms that have plagued other charter initiatives, a requirement for demonstrated improvement in student academic performance in order to maintain charter status must be part of the package.

In states with existing charter laws, the new legislation should not attempt to strengthen these laws in areas of perceived weakness, beyond lifting limitations placed on who can apply for a charter. Objective One must be to extend chartering opportunities to every district, if it chooses to exercise them. Any bill that includes this provision will meet opposition from teachers' unions, colleges of education, and from some who see the charter movement and anything associated with it as a child of political conservatism. Each change the bill attempts to make will attract new opponents, and a common strategy of opponents of a piece of legislation is to load it down with amendments that attract additional opposition. Limiting the changes to those necessary to allow every public district to apply for charter status and keeping the bill as "amendment free" as possible keeps opposition to a minimum—though it will most likely be a well organized, vocal, and aggressive minimum.

In states with legislatures that have a conservative majority, the idea might have immediate appeal, and opponents will have a more difficult challenge derailing the proposal. Where there is a more liberal majority, the objective has to be to demonstrate that this approach to education reform conforms with the majority party's agenda by discouraging greater student migration to the private sector, by taking some of the steam out of the "school choice" argument by providing both rigor and choice within a public district, and by preventing the divisions in resources, talent, and student body that creation of a separate charter school within a district can create. The emphasis being placed by the Obama Administration on the spread of charter principles and opportunities, while still protecting the form and structure of public education, should have appeal with more liberal legislatures.

Finding the Right Champions

No matter which party controls state government, those leading the charge for school reform need to find a champion within the majority party and ideally cosponsors for the bill from each party—then begin the process of getting as many legislators as possible from both sides of the aisle to sign on to the bill as cosponsors. As opposition becomes organized and vocal, proponents must be better organized and more vocal. In the United States we greatly underestimate how responsive elected officials are to letters, e-mail messages, twitters, and phone calls, and Frank and Caroline—whose idea this was in the first place—need to encourage everyone they know to support the "Strong Education for a Strong Community" legislation, or whatever the campaign is labeled, by calling their representative.

Why will teachers' unions and colleges of education oppose the bill? Because good charter legislation frees the district of many of the restrictions created by teacher certification laws. If, for example, a school district wishes to hire a PhD chemist who has decided to leave DuPont or Monsanto because she has a passion to teach high school juniors, in many states she cannot be hired without first completing a series of teacher education courses that cover classroom management and teaching methods—courses that are the lifeblood of university teacher education programs. Despite the reality that most faculty members at the same university have not had these methods courses, they are expected of those teaching at the elementary and secondary grades, unless freed of these requirements by a charter law. Teachers' unions often see certification as one of their important protection mechanisms, and naturally oppose efforts to bypass it. Champions of the legislation must anticipate this opposition and prepare and distribute documentation that shows that many of the teachers in highly effective charter schools are not certified by the traditional standards but come out of programs such as *Teach for America*, and that little relationship can be shown to exist between completion of this prescribed set of methods courses and effective teaching.

In a report calling for a national framework for evidence of effectiveness of teacher education programs, the American Association of Colleges and Universities noted:

> For many years, administrators of teacher education programs focused on inputs, best illustrated by the almost exclusive attention in

accreditation to such input measures as the quality of faculty, nature of the curriculum, adequacy of the budget, and the like. Administrators, to be sure, aimed to produce good teachers, but a focus on outcomes was absent. And the definition of good teachers was limited, usually being defined as teachers whom principals said were good or fit in well. In this new age of accountability, however, a focus on inputs is simply inadequate. The key measure of success for teacher education programs today must be how well they produce teachers who can demonstrate that they can produce learning gains in K-12 pupils.[2]

While this report presents several models that are being developed to assess the effectiveness of teacher preparation, none of the evidence called for by the report is provided, and considerable space is committed to explaining why coming up with that evidence is so challenging. Evidence *can* be provided, however, demonstrating that teachers prepared using alternative methods are doing an excellent job in the classroom, with impressive results in student performance. For legislation to pass that extends chartering privileges to all districts, its proponents will need to be armed with information demonstrating this teacher effectiveness and must repeatedly reinforce two central messages: that "districts have a choice about whether to participate," and "if they choose to elect to follow the charter path, districts will be held to strict performance accountability."

The key to enacting legislation is to remember that in most cases a bill does not pass because the majority of the population supports it, or are even aware of it—or because most legislators understand and embrace it. Legislation passes because the *right* politicians want it to pass and are willing to throw their considerable influence behind it, call in political favors, and negotiate trades with others who have their own set of special interests. This same group of legislators can keep the bill from passing, despite broad public support, and if a district and CORE Team lack connections with this power structure, it would be worth the effort to learn who they are and to work with the districts in which they live to build an effective political coalition before pushing legislation forward.

It is important before wrapping this chapter up to stress that extension of charter privileges to school districts, though a significant boon to district reform, is not absolutely essential to bring about significant improvement. As Leopold in Missouri, Ysleta in El Paso, and Montgomery County in Virginia have demonstrated, communities and school boards who have committed themselves to creating a culture of education can change the way both students and their families think about education. Start the reform process, demonstrate that the district is serious about improving student achievement, then push for legislation that will enable the district to add those extra elements that are making our best charter schools models of excellence and can do the same for every committed district.

CHAPTER FIFTEEN

Managing with Data

"The plural of an anecdote is data."
—Attributed to Nobel Laureate, George Stigler

One of the most successful reform efforts in education in recent years has been a community college initiative initially sponsored by the Lumina Foundation called *Achieving the Dream*. The primary focus of the initiative has been to close the educational performance gap between students from disadvantaged backgrounds and the rest of the student population. Twenty-seven colleges in five states were initial participants, and in order to receive support from the Foundation, they were required to meet a number of expectations clearly stipulated in the application. The most important of these was that every college had to develop a comprehensive database that tracked each student's academic progress from year to year on a cohort basis, with these data reported to a central information manager and shared among the participants in the initiative. In this case, "cohort basis" meant that as each class entered the college, its students became a "data set," and each student in that cohort was followed to see what happened to both the student and to the cohort group in terms of academic progress.

The second expectation of *Achieving the Dream* was that every college, based on what its data showed, was required to design two or three strategies to improve student achievement. The strategies were left largely to the colleges to determine and design, but the success of the strategies had to be tracked using specific quantitative measures, with this assessment information shared across the initiative. Over time, the program has grown to include over 100 colleges in close to half of the states, with a dozen different foundation groups or state agencies assisting with funding. The result has been a large database against which colleges can measure performance, and a tool kit of tested strategies from which institutions can draw ideas when looking for solutions to specific academic challenges.[1] As a vehicle for education reform, *Achieving the Dream* has demonstrated twelve data-related principles that apply equally to every level of education and must be part of rescuing America's elementary and secondary schools as well. This chapter discusses each, with specific application to K–12 education.

> **Principle 1:** If you don't know how well you are doing in specific discipline areas, you don't know where improvement is needed.

Despite its recent brush with quality control issues, one of the most studied and admired "quality improvement" systems in the world is the Toyota production system that produces an automobile considered a little better with each production year, and always rated among the top performing vehicles. For decades, Toyota set the standard for automotive quality control. Part of this impressive system is what the company calls *jidoka*, a term that traces its origins to an automated loom invented by Toyota's founder, Sakichi Toyoda. To insure that the loom did not continue to produce cloth if a thread broke or the automatic shuttle changer malfunctioned, Toyoda built into the loom an automatic stop device that brought production to a halt whenever a problem was detected. Toyota Motors interprets the term *jidoka* to mean "automation with a human touch," and empowers every employee to stop the process if a problem or defect is identified, by pulling what is called the andon cord. Production stops, the reason for the defect is diagnosed and changes made to keep it from happening again, and production continues—resulting in a continuously improved process and product.[2] Key to the Toyota philosophy is that the company expects every employee to pass along to the next employee in the process a product that is completely ready for the next set in production.

I recall hearing a radio interview with a Toyota production manager in which he spoke about plant production meetings as focusing on "bad news first," stating that as the manager of each system reported on how his or her area was doing, the initial emphasis was on what was not working as well as it might, and what was being done to make it better. A Toyota plant Web site notes that over 90,000 employee suggestions are adopted each year, leading to a constant flow of process improvements.[3]

It is not reasonable for every teacher in our public schools to pull the andon cord and bring production to a halt every time some student fails to turn in an assignment or does poorly on a test. But it is possible for the process to be monitored with such regularity that the cord can be pulled as soon as a term ends if students are not showing appropriate levels of achievement. All teaching would not halt, but a quality improvement team could be assembled to see what changes can be made that will improve results before the end of the next term. The problem we have in most of our schools is that we do not know from term to term, and sometimes from year to year, how students are doing relative to what is needed to progress successfully into the next grade, because testing is sporadic, is not comprehensive, and often goes unreviewed.

In chapter 10 on teacher evaluation, I argued that every teacher should insist on standardized testing at every grade, and upon a progressive and carefully aligned curriculum from grade to grade within schools. This standardized assessment within the district should measure on a regular basis how successfully students move along this achievement continuum, with quarterly quality assurance meetings at which performance data are laid out on the table, progress is evaluated in each critical learning area, and problems identified and addressed. Fair and objective teacher evaluation depends on measuring student progress while under the tutelage of *that* teacher, and without consistent, accurate, and regular data, this can't happen.

In one of the colleges participating in *Achieving the Dream*, a quality assurance meeting in the English department found that students in English I were doing significantly less well every alternate year. Two teachers were alternating annually between English I and English II, and when Teacher A was instructing in English I, student achievement and retention were significantly lower than when Teacher B taught the course. Both teachers had equally positive results with English II, and during the meeting discussion Teacher A admitted that he really disliked teaching English I, but loved English II, and he suspected that his dislike for the basic course showed in how students performed. Teacher B enjoyed both, and was more than happy to accept a schedule that kept her with English I each year. A change was made in scheduling and by the end of the next year—a year that should have been Teacher A's with English I—student achievement had improved markedly. This simple change and the resulting affect on student achievement would not have occurred without good data honestly evaluated on a regular basis.

Principle 2: When change occurs in the curriculum, it should be data-informed.

As chapter 7 on math reform noted, one of seven commercially produced math curricula is used by over 90 percent of school districts in the United States, and a constant battle rages over which of these math programs is most successful. A common complaint from elementary and secondary teachers is that districts respond to these battles and to the sales pitches of publishing representatives by jumping from one curriculum to another whenever math performance doesn't meet expectations. As a result, teachers are constantly revising lesson plans, learning new approaches, and scrapping carefully developed teaching aids that took hours or days to prepare. In discipline areas such as reading, two or more methods may be used within the same district at once, as teachers attend conferences or read reports that convince them that something will work better than the program being used in the lower grade.

The truth is that, in most cases, these changes occur with no objective data that inform those making the decision about why the old approach is not working, or why another might be better. If students are doing poorly, no systematic assessment determines whether it is because the curriculum is inadequate, the teachers are poorly trained in its use, greater time-on-task is needed, or students are not getting the support they need at home to complete homework and prepare for class. If one of these latter reasons is to blame, no change in curriculum will make a significant difference. The *Achieving the Dream* colleges have found that one of the most important functions on their campuses is that of the institutional researcher or research office—a function that in K–12 school districts is often shared with other responsibilities. An important element of reform in every school is creation of a research office that tracks performance in every classroom, provides each grade with understandable data to measure student success, and compares performance with whatever standard the school has established as its benchmark.

The tendency to change curriculum when performance is substandard is a reflection of hesitance to examine student performance in critical ways on a class-by-class and teacher-by-teacher basis. If two-third grade classes are doing poorly in reading and two are not, and all are using the same reading program, the curriculum either is adequate and poorly presented by two of the teachers, or is inadequate and can be supplemented by innovative teachers to get the job done. Without additional training for those whose classes are doing poorly, a new reading program will probably produce similar results. Data gathered and evaluated over a period of years can help a district determine whether a teacher is struggling, a particularly strong or weak student group is working its way through the system, socioeconomic conditions are changing, specific social and cultural challenges in one region of the district are influencing performance, or if other distractions in a class may be creating disruptions. A change without good data is no more than guesswork.

Teachers are likely to complain that there are so many of these factors that influence student performance that basing teacher evaluations on student success is unfair. But, as noted earlier, the opposite it true. Keeping careful assessment of student progress out of the evaluation process leaves schools with no sound basis for taking corrective action and may result in blaming teachers who are performing well, as easily as it may protect them from blame. Evaluation without data has been one of the fatal flaws of our educational system and one of the greatest contributors to its decline. If a teacher's job is to aid those in his or her charge with acquisition of a prescribed body of knowledge and a specific skill set, how else does a school adequately evaluate teacher success if not by measuring whether the knowledge and skills were acquired? By assessing student behavior? By checking the degree of interaction with students? By grading the teacher's presentation style and warmth? All of these could be characteristics of a good day-care provider, and are no indication at all of whether anything was learned.

Teacher assessment without student performance information is both unfair and ineffective.

Principle 3: Careful performance tracking must be a condition of reform support from the State—in this case, state permission to have greater freedoms and flexibility.

There may appear to be an inconsistency in saying, on one hand, that school reform will not succeed if left to state departments of education, while insisting, on the other hand, that these departments have a vital role to play. But that is exactly how reform can best be enacted. States must grant permission to schools to revise curricula, select faculty, and change school calendars with greater autonomy, but must insist they demonstrate on an annual basis that student performance is either improving in measurable ways, or has reached standards acceptable to both the community's and the state's assessment of what an internationally competitive school should achieve. State support must be tied not to processes, not to certifications, and not to some prescribed calendar, but to student success results using assessment tools that are sufficiently common and well defined that they can't be manipulated by excluding students from testing or by using a locally developed instrument that has questionable validity. These state assessments do not need to be at every grade level, but should be frequent enough that schools recognize that progress must be maintained from year to year if students are to meet required achievement goals. In years and for grades in which the state does not assess, the school must do so.

One of the principal keys to the success of the *Achieving the Dream* initiative is that the participating colleges have broad autonomy but strict accountability. Funding support is tied to achieving specified performance objectives. For reform to succeed in the K-12 sector, states must extend these same freedoms to schools to reform, but must tie the financial reward system to clear demonstrations of student achievement if this flexibility is to be allowed. Schools that do not achieve should be placed again under the controls of strict state oversight, and should lose incentive funding that is tied to student achievement.

Principle 4: Schools must be completely open with their data, holding regular internal discussions about what these data mean and where they are directing change.

At Patrick Henry Community College in Martinsville, Virginia, one of the lead institutions in the *Achieving the Dream* initiative, once or twice a year the college holds a "Data Summit." All faculty and staff come together for a daylong review of how the college is doing related to a set of key performance indicators. The Institutional Researcher runs through a course-by-course analysis of student achievement data in the area being examined, noting where the college and its students are making progress and where there are weaknesses in achievement. It is essentially Toyota's production review meeting, where key areas are reviewed for needed improvement, and faculty participate in an open discussion about how changes can be made to produce better results. At each of these summits, teachers who have been particularly successful at improving student performance discuss what they are doing, and why it works.

In addition to holding data summits, Phillips Community College in Arkansas has an intranet Web site where student achievement data is regularly posted for internal review and discussion. The philosophy is one of "complete transparency," with everyone knowing how the college is doing in areas of academic achievement, and continuous discussion occurring about how deficient areas can be improved.

Both of these colleges are committed to the principle that data are neither good nor bad, but are simply an indication of what is happening within the system. They prefer the expression "data-informed" to "data-driven," and use their data to guide, rather than drive, change. Like the instrument panel on an airplane, the data indicate the direction the plane is headed, its speed, if engines are running efficiently, whether in a climb or a descent, and when in a steep turn. The irony of tracking school performance has been that, while we would never dream of flying in a plane if we knew the pilot and crew were ignoring these indicators, we have chosen to fly our schools by the "seat of our pants"—never too certain if we are climbing or descending and at what speed, and occasionally uncertain about the direction we are heading.

One of the leadership principles espoused by Mary Parker Follett's Law of the Situation, discussed in chapter 5, is that although the answer to every problem exists somewhere within the problem, we are most likely to discover it when we lay all of our cards on the table and bring the talents and experience of the whole organization to bear on the issue. If a school restricts access to its data and chooses to limit its analysis to a select few, it also greatly limits the probability that the best solutions will be found. When a school wants to generate the best ideas for how to improve student success, it needs to get its data out in clear view and let everyone see it, think about it, and offer recommendations.

Principle 5: To be useful, data must tell us about the performance of specific students and cohort groups—not aggregates.

My family arrived in Iran a day or two before my ninth grade school year began, so my first days in class were also my first days in the country. Each student at the American Dependents' School where I enrolled was required to take a class in basic Farsi, the dominant Persian language, and one of my first classes on my first day was in Farsi. The teacher, a severe Iranian woman, began the class by asking each student to say some simple phrases in the language, and selected me as one of her first victims.

"How do you say 'How are you,' or 'how is your health,'" she asked?

"I don't know," I said, and before I could tell her that I had just arrived in the country the day before, she launched into a lengthy lecture to the class about how disrespectful it was of Americans to come to a country and not learn even the rudiments of the language. Although there was some truth in what she said, it was not true of this particular group of students, and I learned quickly that most of my classmates spoke pretty fair conversational Farsi. But the teacher was making a common and often repeated mistake of making an academic judgment based on the performance of an outlier—a person who was not representative of the group she was interested in assessing.

Every year I see two common examples of how data can be misused in this way, and in both cases it is by ignoring the tracking of specific students and cohorts. The first appears in the form of announcements by states or school districts that they have a specific percentage of high school students who are going directly on to college. The rate is often around 60 percent, and sounds good on the surface. What this statistic usually doesn't tell us is 60 percent of what? It usually means 60 percent of those who graduated from high school the previous spring—not the number who entered the high school four years earlier. In fact, as was illustrated earlier, if we were to track that ninth grade class student by student from the time it commenced secondary school, nearly a third of its members on average didn't even graduate with the class. It is 60 percent of the 68 percent who finished high school on time who actually went directly on to college—closer to 40 percent of the class.

And the 68 percent who graduated with their class may not have represented accurately the ninth grade persistence rate, because the school will announce the *number* who graduated on time, not the percentage of the students who actually entered that secondary school as ninth graders. As high school students moved into the district, they were added into the class numbers to replace students who dropped out, often inflating the graduation rate of the original cohort. A critical part of being open and transparent about data is that the numbers we are open about need to be an accurate reflection of how well individual students and consistently identified student groups are doing.

This approach to tracking performance becomes particularly important when assessing areas such as math achievement from year to year. If students who complete Algebra I in the eighth grade are not the same ones who complete Algebra II at the end of the ninth grade, there is no accurate assessment of how well their eighth grade year prepared them unless they are assessed at the end of the eighth grade, and the same students are tracked through the following year. In a class of thirty, if four or five students move into the district between the eighth and ninth grade years, their performance can dramatically skew the achievement data for the ninth grade class—either positively or negatively. A curriculum revision decision based on poor ninth grade performance—or an assumption that it is doing especially well—may be based on the performance of students who were not even with the district during the critical pre-Algebra and Algebra I years. This doesn't excuse the district from bringing weak transfer students up to standard, but these students cannot be used to determine how well the standard is being met by teachers at earlier levels. Schools must carefully define the cohort groups they track and keep these groups intact as they follow progress, and they must move outliers into other groups that are similarly tracked based on a common element such as the year they entered the system, or their level of proficiency when they began. It is the academic equivalent of insuring that we are comparing apples to apples.

Principle 6: Not all data are equally useful.

A criticism often leveled against educational institutions in general is that they are data rich, but information poor. States have mandated reports on virtually everything for years—occasionally adding a new requirement but rarely discontinuing one. As a result, both state archives and school mainframes bulge with information, much of which is never examined, and even less of which is particularly useful. Schools are not likely to convince states to stop collecting so much data, but they can make internal judgments about what they will collect and analyze themselves, and how it will be presented. The state may, for example, not at all be concerned about cohort data, meaning that its cumulative information concerning class performance is not especially helpful—requiring the school to keep a parallel set of records that can be manipulated to be sorted according to the criteria the school desires.

Many school districts and state systems are captives of large commercial data management systems that are designed to meet the basic information needs of many schools in many states, but are expensive to modify to meet the more specific needs of a single district. This may necessitate running parallel systems to provide some of the specific information a school needs, or downloading information and manipulating it in Excel or some other database—all the more reason to have a dedicated institutional researcher on staff.

Principle 7: Data need to be used to help us understand not only what is happening, but why.

For the sake of illustration, let us say that Central Suburban High School (CSH), the single high school in a fictitious district in the Southeast, receives students from three middle schools that cover grades six, seven, and eight and represent widely diverse populations. As CSH reviews its testing performance for a given year, it discovers that there has been a precipitous drop in ninth grade performance in language arts, particularly in written English. The year before, a popular and very successful English teacher retired after twenty-seven years of service, and was replaced by a first-year teacher who just completed a master's degree in English Literature at a nearby university. As the "new kid on the block," this teacher was assigned most of the sections of ninth grade English. The conversation between the Assistant Superintendent for Instruction, the high school principal, and the Chair of the English Department goes something like this:

Assist. Super: "This drop in ninth grade performance in Language Arts is disturbing. What do you attribute it to?"

Principal: "I'll defer to Mrs. Saunders on that. What do you think, Pat?"

English Chair: "As you know, we had a staff change in a critical position there this year. We've evaluated the new teacher during the year and his classroom performance seemed to be pretty good. But apparently the students aren't getting the material."

Assist. Super: "Where do we go from here?"

Principal: "We'll call him in and talk to him. The biggest drops seem to be in basic areas, such as vocabulary and English usage. He's a Lit graduate and may be spending too much time on literature and not enough on fundamentals. He's still on probation so we'll give him a year to show some improvement in test scores, then will have to consider other options."

There is an immediate assumption in this conversation that the change in performance is a reflection of a change in teacher and that the problem lies with the new instructor—an easy judgment to make, given the drop in achievement that immediately followed the retirement of a star faculty member. But the conversation violates one of the fundamental rules of assessment and evaluation—a new Fundamental Law of the Universe for evaluating raw data. *An apparent relationship does not necessarily demonstrate causality.* Stated more simply, just because two things seem to be related, it doesn't mean one caused the other to happen.

Suppose, for the sake of our example, that performance on the ninth grade assessment is broken down by the middle schools from which the students transferred to the high school, and it is found that students from two of the schools (Schools A and B) did *better* than previous students had done on the test. But students from the third school, School C, did significantly worse, resulting in the dramatic decline in average scores. This would indicate that the problem is not with the new teacher at all, but that something is happening at Middle School C. Were there teaching changes at School C two years ago that might have affected seventh and eighth grade learning? Did major flooding close school for a week, shortening the time eighth graders at School C were able to spend on critical material? Did a new poultry plant open in the area served by School C, introducing a large population of students for whom English is a second language? (Cohort tracking would show that this new student group appeared in the district at all levels, and that a few ninth graders were included in the testing who did not come from any of the middle schools in the district.)

Through this process of what is sometimes called "drilling down" through the data, a skilled data manager can identify not only what has happened but also why the change is showing up—saving the district the possible loss of a talented new teacher and the probability that the wrong "problem" will be addressed. There is also a general observation in research that quantitative (statistical) studies are often quite good at telling us what is happening, but not too successful at telling us why. To get to the "why," another approach to research might be needed.

Suppose we find that, in our example above, not all students coming into the ninth grade from School C have done poorly, and in fact that there is a clear ethnic divide in performance with students of Hispanic background doing the least well. Through a focus group, a small group interview with eight to ten of these students conducted by a trusted Hispanic community leader, the students explain that many of their families are transitioning from labor jobs into small, family-run businesses—restaurants, a grocery store, a yard care service, an after-hours cleaning business, and a car detailing service. Every member of these families who have started entrepreneurial enterprises works to support the family business, and as soon as students leave after class they go to work for the next six hours to help the fledgling businesses succeed. Practically no time is left for homework. Without this information, the school is again likely to jump to the wrong conclusions, sanctioning a teacher, changing a curriculum, or making judgments about a student group that would be completely unfounded. In making important educational decisions, it is critical to know not only what, but why.

Principle 8: Sharing data among schools stimulates ideas and improvements.

In the extreme southwest corner of Virginia, a group of school districts has cooperated to form three Math Collaboratives that bring high school, middle school, and elementary teachers together several times a year to discuss what is working well in math education, and where they are having problems. For the collaboratives to function well, two conditions are essential. The schools must be willing to share information about how students are performing, and teachers have to set egos aside and talk candidly about where they are having difficulties and need help. This willingness requires a belief that within states and regions education is not a competitive enterprise, and that everyone benefits if each school improves. In the case of the Virginia collaboratives, a math faculty member from a nearby university acts as facilitator, coordinating meetings, asking probing questions, and offering resource support. In addition to exchanging ideas and offering suggestions, the group has found that one of the unanticipated benefits of the collaborative is that many of the students who change residences during the year move within the southwest region of the state, showing up in one of the other collaborative schools. The greater the coordination and alignment of curriculum among schools, the less time teachers have to commit to bringing new students up to grade level as they move from school to school.

Achieving the Dream holds an annual strategy forum at which colleges that are making significant improvements in key curriculum areas or who have developed particularly useful models for assessing performance present their successes to colleagues. Through this exchange there is a constant dissemination of "best practices" throughout the network of participating schools—a model that could easily be emulated as a number of school districts within a state or region undertake the reform process. I was surprised while collecting information about "best practice" schools for this book to discover how hesitant even successful schools are to share their data. There is a general air of suspicion about making what is clearly "public information" available for public scrutiny. In one case I called and e-mailed an elementary principal five or six times whose school had been referred as a model of language immersion, without response. The person who answered the phone at another large district immediately wanted to know if I was a reporter, then, after a long delay, informed me that the district's public information person was not available. Could I please send a detailed written request for the specific information I would like to see? I did, and it was neither acknowledged nor the data provided. In both of these cases, my expressed interest was to feature the schools successes, but suspicion ruled the day.

In most instances, to the degree that data on school performance are available, they are a matter of public record, and a citizen can find them on a state Web site or insist that they be provided by the school. A provision of any legislation more broadly extending chartering privileges to school districts should include a requirement that data related to a key set of performance indicators be routinely posted on a publicly accessible Web site. Then, those who are supporting the school with their taxes and donations can see what they are getting for their investment, and districts can compare themselves with peers without the cost and complication of extensive reporting and collecting networks. Just as teachers cannot fairly be evaluated without pertinent and accurate performance information, neither can districts as a whole.

Principle 9: Focus on curriculum change leads to greater improvement than does focus on student support programs.

One of the most important findings of the colleges that are participating in the *Achieving the Dream* initiative is that while changes in support services such as advising, tutoring, and mentoring can lead to some increases in performance, in order for dramatic change to occur, there must be a change in curriculum or teaching approach. Returning to Kay McClenney's Fundamental Law of the Universe, schools must be willing to admit that the assignments, courses, and programs now in place are designed to produce exactly the results they are now producing. So, tutoring, counseling, and mentoring may help students do what they are currently being asked to do a little better, but we are not asking enough—so we will never accomplish what needs to be accomplished without revising content. Continuous and systematic data analysis tell us where content needs to be revised and strengthened, and education will never make the changes necessary to bring about significant reform without bold, data-informed curriculum revision.

Principle 10: There is value in having an external coach serve as an objective advisor and critical friend.

One of the most useful developments of the Lumina Foundation's reform initiative with community colleges was an early requirement that each college work with two external "coaches," one trained in data analysis and the other with a background in organizational leadership and change management. These coaches are advised to resist working as consultants, presenting their own "solutions" for the colleges' problems, but to serve as "critical friends," helping the schools organize and review data, identify areas of weakness, and ask the difficult questions about what is needed to achieve desired ends. An underlying assumption is that the school as a collective whole has a better understanding of its students and the community than do the coaches, but may need assistance in raising and facing the most difficult questions and sustaining the momentum needed to support a change agenda. The coaches serve three useful purposes: (1) they bring in fresh sets of eyes that are not blinded by the internal culture and assumptions of the institution; (2) they tie the school to a broad network of external links and resources that can direct it to others who are struggling with similar challenges; and (3) they can probe and challenge with relative freedom, since they fall outside of the personnel structure of the organization. They often offer an additional benefit by providing personal expertise that is missing within the school system. This becomes particularly important if the school does not have someone with a strong data analysis background and the coaches are able to help the district think about what information will be useful, how it can be retrieved and presented to faculty and staff in an understandable way, and how the "drill down" process can help the district isolate major contributing factors to identified problems.

Support for most of the coaching activity is provided through grants from one of a number of funders—but not all. Several colleges in the initiative have chosen to self-fund their participation in the initiative, anxious to take advantage of the resource pool available to members and to be part of "the movement." They see value in drawing energy from the excitement and enthusiasm of others who are engaged in the same work. They also find that, as they retain more students, their improvement efforts pay for themselves. As school districts begin to pursue a reform agenda, they might reasonably form their own initiative group and provide coaching assistance to each other to bring in those fresh sets of eyes and an external voice that can raise the difficult, but often unspoken, questions.

Principle 11: A central repository for data allows benchmarking and progress comparisons.

As one of the principle partners in *Achieving the Dream*, the American Association of Community Colleges agreed to serve as a data repository for information submitted by the participating institutions on key performance indicators, and committed some of its research staff to organize the data into what it calls "data cubes," which allow colleges to sort and manipulate their own information to examine changes, trends, and cohort achievement. The database is also designed to help schools identify peer institutions with similar demographics, student body size and characteristics, and academic challenges. Each participating college provides the web organizers with a list of its reform strategies and submits annual updates on how well activities are working.

Two other national community college networks formed outside of the *Achieving the Dream* project provide examples of how shared data sets can help institutions benchmark themselves against those who excel in key performance areas, and share information about best practices. The Community College Survey of Student Engagement, recently renamed the Center for Community College Student Engagement, provides instruments that allow member institutions to evaluate the effectiveness with which students are engaged with their school environment, and assesses other student involvements that may affect performance. Two-thirds of the nation's community colleges participate in the survey, providing a vast network of shared information and resources, and a benchmarking system that enables each member to compare student responses to selected peer colleges.[4] A similar benchmarking network, but one based on a set of key performance indicators identified by member colleges, has been created through the National Community College Benchmark Project, through which members report outcome and effectiveness data in critical performance areas and are able to compare results with those of other colleges. Institutional identities are not provided to other member institutions unless used as best-practice examples, and then only with permission of the featured school. But peer comparisons are made possible through detailed descriptions of college demographics.[5]

None of these collaborative examples utilizes a state or federal database, and in each case the data collected and shared have been determined and modified over time with broad participant input. For community-based education reform to grow and thrive as a national movement, similar cooperative benchmarking initiatives will need to be organized, with members determining what data is to be collected and how it can be shared to provide greatest assistance in improving student achievement.

Principle 12: A relatively small amount of incentive money can lead to significant improvement.

The initial group of twenty-seven *Achieving the Dream* colleges, whether large or small, received a first-year planning grant of $50,000, with support for four successive implementation years at the $100,000 a year level. As other funding sponsors signed on, annual support grant for new member colleges varied considerably, but never exceeded this initial amount. A number of colleges asked to join the initiative as self-funders, anxious to take advantage of the networking benefits and the invaluable data that was flowing from dozens of institutional activities designed to improve performance. One participating president who chose to self-fund and has a background in accounting commented to me that she couldn't afford to be out of the initiative. "I don't have to retain many students to pay for this investment," she said, and within two years of entering the initiative was able to demonstrate that she was covering her costs through improved student persistence.

Good data help a school understand and demonstrate how enhancing student performance and retaining actively engaged students in greater numbers make financial sense. Good data allow for efficient facility utilization, and justify investment in curriculum revision, improved pay schedules, and faculty development. How much money a district has to spend per pupil appears to have a much smaller influence on student achievement than how the district chooses to spend it, and without clear data, openly reviewed and objectively analyzed, these distinctions can't be made.

Barriers to Effective Data Use

A study by the Rand Corporation found that a common series of factors explain why many educators, both teachers and administrators, fail to use data effectively. These included inaccessibility to the information, concern about its quality and reliability, lack of incentive to be data-informed, timeliness of important information, personal time available to review information, staff support, and pressures created by mechanisms such as "pacing plans" that reduce personal flexibility and prerogatives. In many cases, a culture of inquiry is not encouraged by school leadership, and encouragement by states to examine the organization in critical ways and implement a continuous improvement plan based on data is not sufficient to overcome poorly motivated leadership.[6]

The experience of the *Achieving the Dream* initiative has been that this last factor is indispensable. When leadership lacks a commitment to data-informed planning and reform and to creating a culture of inquiry, it will not happen. Data will not be disseminated as they need to be, they will not be timely, there will be limited concern about accuracy and reliability, and no accommodation will be made for the time and staff support needed to analyze and use data effectively. A healthy approach to creating a culture of inquiry is to assume and promote the philosophy that when performance is substandard it is more often a *process* problem than a *people* problem. Teachers do not come to school each day thinking, "Today I will do less than my best, and I don't care about the quality of my work." They do, however, come poorly trained, underprepared, under-supported, and ill-informed about where problems exist. A culture of inquiry allows the institution to determine where processes are breaking down, and provides those responsible for process improvement with the tools needed to make adjustments.

Until we have leadership that routinely provides for these requirements, community activists and those committed to education reform will need to serve as institutional coaches and critical friends, asking the questions that will drive necessary inquiry and informed change. "We don't know" is not an acceptable explanation for why students are not succeeding, and an alternative answer that cannot be supported with evidence is equally unacceptable. In every district we need to know how each student is doing and when not doing well, where problems exist and what the probable causes are. We can only address our challenges with wisdom when we have clear and reliable information about what causes them.

CHAPTER SIXTEEN

Building Successful Partnerships

"If you want to be incrementally better: Be competitive.
If you want to be exponentially better: Be cooperative."
—Anonymous

When students at Cottage Grove High School in Oregon's South Lane School District enter their junior year, those interested in a career as an Emergency Medical Technician can enter a program offered on their high school campus through Lane Community College, located twenty miles north in Eugene. This ten-hour program, scheduled during the first two hours of the day, provides completers with basic EMT certification and can be transferred after high school into Lane's Associate of Applied Sciences EMT program, leading to an opportunity to sit for Oregon's EMT-Paramedic certification exam.

In Dallas, two charter high schools located on the campus of Richland Community College accept 300 juniors and 300 seniors annually into the Collegiate High School of Math, Science, and Engineering, and another 150 into the Collegiate High School of Visual, Performing, and Digital Arts. Tuition and books are free for those accepted, and students participate in a "rigorous academic experience" leading to both high school graduation and completion of an associate of arts degree from Richland.

Rather than bring high school onto its campus, The Community College of Baltimore County takes its freshman placement exam into school districts across its region and administers it to students as they complete their sophomore year. If the assessment shows deficiencies in math, reading, or English that would prevent students from entering regular credit courses at the college upon high school completion, counselors direct them to additional coursework during the junior and senior years to adequately prepare them for college-level work.

These colleges are among hundreds across the United States who are responding to the call from President Barak Obama for community colleges to provide Americans "a chance to learn the skills and knowledge necessary to compete for the jobs of the future."[1] For most in community college circles, this charge is no more than an echo from the past. Over half a century ago as thousands of GIs returned from World War II, President Harry Truman assembled a commission of school, civic, and business leaders to recommend new direction for education in postwar America. It was this 1947 Truman Commission that coined the name "community college," and recommended:

> The time has come to make education through the fourteenth grade available in the same way that high school education is now available. This means that tuition-free education should be available in public institutions to all youth for the traditional freshman and sophomore years or for the traditional 2-year junior college course. To achieve this, it will be necessary to develop much more extensively than at present such opportunities as are now provided in local communities by the 2-year junior college, community institute, community college, or institute of arts and sciences. The name used does not matter, though community college seems to describe these schools best; the important thing is that the services they perform be recognized and vastly extended.[2]

During the 1960s, these new community-responsive postsecondary institutions were created at the rate of one per week, and now nearly 1200 of them blanket the United States. This uniquely American invention, now being copied by countries around the world, provides a bridge between secondary school and what was once a university education accessible only to the wealthy or to an intellectual elite. Created at a time when America realized that all citizens would benefit from more education, most have become what are referred to as "comprehensive" colleges, offering an array of services that include freshman and sophomore baccalaureate education, career and technical training, precollegiate refresher education for adults, customized training for business and industry, and continuing educational opportunities for those who simply wish to enrich their lives or improve personal skills.

In most respects, the divisions we have created among and between layers of education in the United States are artificial. Colleges developed as our first formal education institutions to provide training in law, philosophy, classical languages, and theology to the homeschooled sons of our colonial gentry. As public elementary and secondary schools developed, they initially prepared this same privileged class for university study, and trained those who would teach the next generation of statesmen, lawyers, and theologians. But the advent of community-based colleges that extended local education to a thirteenth and fourteenth year suggested that what had previously been distinct divisions between secondary and postsecondary education and between class-based opportunities were blurring, and that a community's education providers should be thought of as a resource available to, and shared by, all.

Shared K-12 and Community College Resources

The earliest manifestations of K-12 and community college collaboration appeared as high school buildings made classrooms available for the first two years of baccalaureate instruction, with faculty shared across educational levels. In recent decades, this has evolved into dual credit or concurrent enrollment programs through which students gain college credit while still in high school, or take classes that apply to both degrees. In 2009, all but three states offered dual enrollment programs, presenting an immediate way for school districts and nearby colleges to share enrichment initiatives.[3] Smaller districts that are unable to offer two semesters of calculus or advanced physics partner with the area college to make these courses available, and when expanding foreign language opportunities, high schools may choose to make faculty available to the college in French or Spanish, with the college sharing a specialist in Chinese or Arabic. Richland Community College offers twelve languages, including Arabic, Chinese, Japanese, Farsi, and Russian, and when nearby high schools are unable to afford this variety, it makes sense that the college's resources be extended to students interested in these languages.

In career and technical areas, the large commercial producers of computer networking solutions such as Cisco and Microsoft now commonly offer both high school and community college training programs, covering exactly the same material at the beginning levels. In a number of cases, institutions choose to offer a single unified program that avoids duplication and moves interested students more quickly along this career path. Jointly administered career centers serve as dual enrollment institutes for high school and college students alike who have chosen to pursue a career pathway—with certification depending less on years in school than on demonstrated mastery of key learning objectives.

Much of what schools do to prepare teachers and enhance instructional skills is not grade-specific, and faculty at every level can collaborate to learn about new technology applications or cooperative learning techniques. Richard DuFour, a former Illinois superintendent, writes eloquently about the power of professional learning communities where faculty come together across grade levels and disciplines to discuss learning strategies, compare results, and coordinate curricula. To function most effectively, DuFour maintains that these communities must focus on three "Big Ideas:" that our mission is not *teaching*, but *student learning*; that student learning is accomplished most successfully through developing a culture of collaboration; and that the effectiveness of these collaborations must be judged by the student learning results they produce.[4]

California's Partnership for Achieving Student Success (Cal-PASS) is such a collaborative and brings together K-12 teachers and postsecondary faculty to review longitudinal data and discuss strategies for improving student success as students move from one institution to the next. A network of 30 four-year universities, 108 community colleges, and more than

5,000 K–12 schools share data compiled by Cal-PASS on more than 235 million student records, including information on demographics, coursework, standardized test scores, and graduation rates. In fifty-five smaller Professional Learning Councils, faculty meet monthly in subject-specific sessions to discuss data that follow local students as they move through the system, working collaboratively to identify problems and evaluate new approaches to improve student learning.[5]

If reform is to be a community-based initiative, it must involve all of the educational resources of the community. The Pew Bridge Project, an initiative that focused on helping schools and colleges improve student transitions, warns that "Reform initiatives at different levels within the entire K–16 education system must be better integrated or the whole mission of increasing opportunities for all students could veer dangerously off course."[6] Universities can serve as an important part of these collaborations if actively involved in the educational life of the region, and there are impressive examples of K–12-university collaborations that address some specific recommendations of this book. Michigan State University's highly regarded Arabic Language program partners with the Dearborn Public School District, for example, to integrate Arabic language learning with classroom instruction from kindergarten through high school, and students graduate proficient in the language and ready to continue Arabic studies at the university if they desire.[7] The College of Education at my own university home, the University of Missouri–St. Louis, partners with schools throughout the region on character education, faculty and administrative development, and academic improvement programs for K–12 students. But the semiautonomous, research-directed nature of many university departments make them less available to collaborative efforts, and often less attuned to the immediate needs and challenges of their elementary and secondary partners. Community colleges provide the advantages of being able to collaborate comfortably in both baccalaureate preparation and career training, with the ability to respond quickly and flexibly to opportunities to share faculty, facilities, students, and program ideas.

The Montgomery County Solution

Maryland's Montgomery College and its surrounding school districts provide a case study in what can happen when education leaders are determined to work together to solve common challenges. When Maryland produced its first set of State Outcome and Achievement Reports (SOAR) in 1994, this community college and the school districts it served in the affluent suburbs north of Washington DC were stunned to find that 57 percent of students entering the college required math remediation, and 40 percent were not reading at college level. Until 1965 the college and the Montgomery County Unified School District had been under a single board of education, and had maintained, after splitting, a cordial but relatively disengaged relationship. With the release of the SOAR data, the County Council asked school leaders to meet to seek common solutions, and through a series of lunch meetings, the college president and local superintendent initiated a plan to assemble college and school staff to review data and find ways to improve performance. The boards of the two systems also began to meet, and agreed to combine databases to allow institutions to examine course sequences that aided or impeded student progress. A jointly commissioned study recommended a series of strategies that, if pursued collaboratively, could improve student achievement.[8]

Data collected through this study revealed that students could follow three distinct pathways through the Montgomery County Public Schools (MCPS): one that included math through precalculus and an honors English course; one that required only trigonometry or intermediate algebra in mathematics and non-honors English as a senior; and one that included geometry as the final math course and senior English that was often below grade level. Almost all students who followed Path One tested into college-level courses when they entered Montgomery College. One in three who followed Path Two required remediation if entering the college, and Path Three led to remediation for virtually every college-bound student. A case study examining the changes that occurred in Montgomery County's educational system notes: "Encouraged by support from board members and local leaders, a bold plan emerged to bring the school system and the college together in a partnership to identify students who were not on track for college readiness and convey to students and parents the urgent need to address this matter. The plan included collaborative intervention strategies to change aspirations and enlist parents in a joint effort to prepare students for college-level work."[9] In a jointly held press conference, the two systems publicly reviewed the data and vowed to work together to find solutions. Together they submitted a cooperatively developed budget to the County, through which the MC/MCPS Partnership now supports over thirty joint projects that it describes as helping to "identify and monitor college readiness through PSAT and assessment testing and support and accelerate student success by easing the transition from high school to college through curriculum development, summer programs, early placement programs, and school-based intervention programs."[10]

The District's Annual Report for 2008 indicated that, over the previous three years, the percentage of students testing at or above the proficient level in mathematics and reading increased in each reported area, in some cases by more than 10 percent. In each case, progress by African American students, the district's largest minority group, exceeded that of the student body as a whole, achieving one of the district's goals of closing performance gaps.[11]

What happened in Montgomery County is less remarkable for the thirty new programs than for what they signify. If we trace this sequence of events, state-produced student achievement data were openly presented to the county's public, prompting community leaders to ask local school officials to *collectively* find solutions. Faculty and staff from both the college and school district examined these data in greater detail and determined that schools were offering an academic path that adequately prepared students for college-level work, but that even those students who elected to go to college immediately after high school frequently chose one of two less demanding paths. Neither of these other options produced similar success, and one virtually guaranteed failure.

Based on these findings, the partners determined that they must test students earlier, inform both students and parents of the consequences of failing to pursue a challenging curriculum, provide stronger and more diverse summer options, share resources that made challenging courses available to more students while in high school, and strengthen professional development for teachers. Boards met to jointly encourage and support change, and the county came forward with additional resource support for innovative solutions. They determined that they could—and must—accomplish what this book recommends: create a sense of urgency in the community, provide local leadership support, hold school officials and teachers accountable, evaluate students and use performance data to strengthen curriculum, involve parents and solicit their involvement, work collaboratively, and marshal whatever resources are needed to energize and support the change needed to create a culture of education.

In the MCPS master plan, the board outlined its commitment to transform school performance, listing as "Critical Questions" those that must continuously guide its actions:

What do students need to know and be able to do?

- How will we know they have learned it?
- What will we do when they haven't?
- What will we do when they already know it?[12]

To insure the greatest likelihood that they can adequately answer these questions, board members established and committed themselves to a set of academic priorities:

- organize and optimize resources for improved academic results
- align rigorous curriculum, delivery of instruction, and assessment for continuous improvement of student achievement
- develop, expand, and deliver literacy-based initiatives from prekindergarten through grade twelve
- develop, pilot, and expand improvements in secondary content, instruction, and programs that support students' active engagement in learning
- use student, staff, school, and system performance data to monitor and improve student achievement

- foster and sustain systems that support and improve employee effectiveness, in partnership with MCPS employee organizations
- strengthen family-school relationships and continue to expand civic, business, and community partnerships that support improved student achievement[13]

The Board's pledge is to do what every board should be doing—work to marshal resources of the entire community to insure that students succeed.

Granted, Montgomery College and its surrounding districts are in the relatively affluent northern suburbs of Washington DC, where many students and their families are not faced with the consuming challenges presented in many urban centers by poverty, violence, and unemployment. But there need be no relationship between affluence and attitude, and between comfort and commitment to cooperate. *Excellence is not a child of affluence.* Many of the neighborhoods that make up the Ysleta Unified School District in El Paso could hardly be described as affluent, and 71 percent of the district's students are classified as "economically disadvantaged."[14] Of the 45,000 attending the district, 19 percent are English language learners. Yet ten of the district's schools have been named National Blue Ribbon Schools. Ysleta's Parkland T-STEM Academy partners with the University of Texas—El Paso, El Paso Community College, and a host of local businesses and community agencies to offer a career choice that it claims will "foster success in college math, science, engineering and technology programs, and ensure high standards and challenging opportunities through integration of theory and best practices."

Partnering is the alchemy that can turn even our basest metals into gold. Martin Luther King once commented that "We may have all come in separate ships, but we're in the same boat now,"[15] a message that seems to have been lost on communities that look enviously at others and blame their educational woes on factors they see as being beyond their control, while resources exist within their communities to accomplish all that needs to be done to create an exceptional learning environment. Districts and their potential partners must be willing to honestly assess weaknesses, seek help from each other when shortcomings are identified, and be prepared to be effective partners.

Elements of Effective Partnerships

Jennifer Brinkerhoff specializes in evaluating partnerships and offers a series of institutional conditions that serve as prerequisites to effective partnering, and also assists with nurturing and maintaining a collaborative relationship, once established. Among the preconditions she lists are tolerance for sharing power, willingness to adapt to meet a partner's needs, commitment to be flexible and responsive as unforeseen circumstances arise, and presence within the partnering organizations of persons of influence who will champion the relationship.

Once established, if the partnership is to flourish, there must be strong support from senior leadership, clear common goals, a sense of confidence and trust in the other's contributions, skills, and capacity, and a belief in a shared constituency, mission, and values. If partners are unwilling to share information, resources, and candid assessments of the success with which the partnership is working, it will falter.[16] Perhaps effective partnerships are such a rare commodity because this seems such a tall order. But Paula Glover found in her study of factors that served as barriers or facilitators of good education partnerships that they can be developed, and are aided by regular interaction between members of the collaborating communities and a sense of common purpose. She found, however, that they will not happen without unwavering leadership commitment, a belief in mutual benefit, and an institutional selflessness that places the success and well-being of students above other agendas.[17]

In reality, elements of successful cooperation apply to *all* partnerships, and significant transformation depends on making a number of them work well: with parents, with local civic and political organizations, and with the business community. Each potential partner has a vested interest in how well students succeed and, if approached with an attitude of respect and acknowledged value and with a compelling discussion of the win-win elements of collaboration, will often surprise us with their generosity and willingness to work on our behalf. Each wishes its contribution to be meaningful and acknowledged when help is received, based on some clearly articulated and justified need. Outstanding schools must become a *community* priority, with all parts contributing to the whole, and each valuing and celebrating the contribution of others.

CHAPTER SEVENTEEN

Sustaining and Enhancing Reform

"There is no power for change greater than a community discovering what it cares about."
—Margaret Wheatley

Each day before I begin work I play a game of FreeCell solitaire on the computer. When working at home, I tell my wife that it gives the machine and my brain a few minutes to warm up—to get all systems up to speed. When in the office, I keep the door closed until the last row of cards zips into place, knowing that passersby will think I'm wasting time and being frivolous if they happen in while a game is in progress. But the ritual has much greater significance to getting my day underway. It demonstrates my own personal Fundamental Law of the Universe—There is always a solution.

I believe wholeheartedly in Mary Follett's Law of the Situation: that within each challenging problem or dilemma lies its answer, if we will only lay our cards on the table and look at them honestly and creatively. The solution is often obscured, and sometimes deeply hidden. But it is in there somewhere, and with enough ingenuity and persistence it can be found. Some who are intimately familiar with FreeCell would argue that this principle should say *virtually always*. There are, in fact, a few deals that are not solvable within the four-box FreeCell format generally used. In the Microsoft Windows 95 version that I generally play, there are 32,000 deals, and according to FreeCell experts, only one cannot be solved—game 11982. In the FreeCell super version called FreeCell Pro, there are 20 million deals, and FreeCell specialist and Pro developer Adrian Ettlinger, using a modification of a computer solver created by Don Woods, determined that, of these millions of games, only 263 were impossible to solve.[1]

That some games can't be played out to a solution seems to negate the parallel between FreeCell and Follett's management principle that the solution always exists within the problem. But in both examples above, the win rate is in excess of 99.99 percent if the game offers four free cells to work with. If the number of cells is increased, every one of these games can be solved. For example, the infamous number 11982 can be solved with five cells. Follett would undoubtedly have said that even the discovery within the four cell format that there is not a solution to some of the hands is, in and of itself, a solution. Thomas Edison is often quoted as having declared, when asked whether he was getting results in his frustrating search to find material for a suitable filament for his light bulb, "Results! Why man, I have gotten a lot of results! I know several thousand things that won't work." In some cases, the solution is that we simply have to figure out a way to change the hand we have been dealt.

Finding the Solution Within

In a training session I attended a number of years ago for a group of manufacturing firms in which Follett's Law of the Situation was discussed, a line manager argued that the cases in which the company had been forced to dismiss personnel could hardly be called "wins." A personnel director quickly jumped in to disagree.

"Assuming that we have done what we reasonably can to help the person succeed," she said, "I've never seen a single case of performance-related dismissal where in the long run, both the company and the individual weren't better off." The win for each in this case came from the recognition that it was time to deal a new hand.

I am hardly a FreeCell aficionado. I simply use it each day as a personal object lesson that, as problems arise, the solution will be in the cards somewhere. I believe the same to be true of the major challenges we face in education. The quote that begins this chapter is from a wonderful little book by Margaret Wheatley titled *Turning To One Another*. The book is basically a collection of poems, stories, essays, and musings about the power of working together to create the world in which we would all like to live, and was written to be a "conversation starter" for efforts to work collaboratively to find solutions. Wheatley observes that "...the world always *only* changes when a few individuals step forward. It doesn't change from leaders or top-down programs or big ambitious plans. It changes when we, everyday people gather in small groups, notice what we care about and take those first steps to change the situation." Her focus is primarily on finding solutions to the destructive consequences of conflict, poverty, famine, and disease, but her observations apply equally to our growing crisis of ignorance. To some, ignorance may seem too strong a word for the state of education in America, but if we think about one of its meanings—to ignore or be unaware of—it is completely appropriate. We are choosing to ignore the dramatic changes that daily reshape our world and our futures, and we seem unaware that our educational shortcomings have reached crisis proportion. They are every bit as much a threat to our nation, its security, and its economic future as were the cold war of the mid-1900s or the War on Terror that seems to absorb all of our attention since the turn of the century.

Wheatley points out that, in times of *recognized* crisis, "people have a great desire to help, so they perform miracles. We discover capacities we didn't know we had. The chaos and urgency of a disaster encourages people to try anything, far beyond any plan or training."[2] But we must first recognize the crisis—not as a few scattered individuals, but as communities. Speaking of the current crisis in education, Shirley Ann Jackson, who presides over Rensselaer Polytechnic Institute, calls it "the quiet crisis," noting that it is "a crisis that could jeopardize the nation's pre-eminence and well-being."[3]

Turning Up the Volume

How, then, do we ramp up the volume on the quiet crisis to the point that it can't be shut out—that it elicits the responses Wheatley describes? How do we reach the point as communities that we "...work intensely together, inventing solutions as needed; we take all kinds of risks; we communicate constantly?"[4] And how do we sustain a reform effort long enough to create permanent change?

John Kotter's tactics for establishing and maintaining a sense of urgency in business settings suggest where communities should start, and begin with "bringing the outside in."[5] Those who lead school reform must become town criers, using every available opportunity to shout "*Oyez, Oyez*," alerting the public to "*Listen! Listen!*" John Rassias observed that "Nothing is real unless it touches something in me and I am aware of it,"[6] and we must use local business leaders, respected clergy, and the leadership of area colleges to frame the message in an immediate and personal way, so that each member of the community feels intimately touched. Following Kotter's advice, we must "send people out" to see firsthand what is happening abroad, and how our best schools in America are preparing students to meet those challenges. We must "bring people in" who understand the nature of the crisis and can present it in terms we understand—who will "bring data in" that demonstrate our relative deficiencies and provide benchmarks for new performance expectations.[7]

To sustain our efforts, we must "behave with urgency every day," acknowledging that we cannot wait for guidance from state or federal agencies, but must begin immediately to work within the statutes now in place, and to change the statutes that limit future opportunities. Reform is not an event, but a continuous engagement with the unacceptable—recognizing that each minute of delay is a minute lost to opportunity. Kotter notes that "urgency begets urgency," and if we expect a continuing sense of urgency in others, we can allow no lapses in our own. And we must confront the NoNos—the people Kotter describes as the "highly skilled urgency killers."[8] There will be dozens of them—possibly hundreds—and they will be passionate, persuasive, and organized. They will argue that the crisis is being overblown—that we have heard these forecasts of gloom and doom for decades, yet international students still clamor to come to the United States because they know we have the greatest education system in the world. If willing to acknowledge any of the data that indicate otherwise, they will insist that proposals are unmanageable, unfair, or impractical, or that we are rushing after solutions without giving them sufficient thought. Delay is the Naysayers' fallback position, and they will attempt to kill virtually any good idea or constructive action by studying it into oblivion or tabling it until next month—and the month after that. Remember the Fundamental Law of the Political Universe, that it is much easier to kill a bill than to pass one—and Naysayers are masters of this law. Kotter describes NoNos as those who "will often do nearly anything to discredit people who are trying to create a sense of urgency. They will do nearly anything to derail processes that attempt to create change."[9] He believes that it is a waste of time to try to win these people over but dangerous to ignore them, and recommends that they be distracted into other pursuits, pushed out of the organization, or exposed for what they are—impediments to making changes that are the only way out of what is a real and imminent crisis.[10]

In the case of education reform, it is unlikely that, once a full-scale initiative is launched to change the community's status quo, the Naysayers can be distracted. Many of them *are* the status quo, and will see their very livelihoods as being at stake. It may be possible to move out some who are sniping from within the organizations, but even if moved out, many will remain in the community and can become even more dangerous when unencumbered by school policy or peer pressure.

Laying the Cards on the Table

The newly appointed and reform-minded president of a college where I did some consulting found when she arrived that enrollment had been stagnant for years, student performance was well below that of similar institutions in the state, faculty were complacent, and program evaluation was virtually nonexistent. But as she tried to move her change agenda forward, she was thwarted at every turn by what appeared to be extremely well-informed community resistance. A supportive staff member finally confided in her that the former president who had run the institution for decades as a personal fiefdom and still lived in the community had dedicated his retirement to protecting his legacy by insuring that things remained as they were when he left. "He presents you as an outsider, trying to mess with the values and traditions that have made the college a centerpiece of community life," the confidant explained. "And he's one of the 'grand old men' of the community. It will be hard to fight him," he warned.

This president chose to expose the resistance for what it was—an attempt to maintain failed programs. Rather than attack the "grand old man" directly, she invited the community to come together to critically examine college performance—demonstrating without pointing fingers that programs that had been lauded in the past as models of success were in fact miserable failures compared to results at peer institutions. Her observation was that "the public is generally brighter than we give them credit for. Given the facts, backed by good, hard data, they resent being lied to. I assure them that it is the very things they value that I am trying to protect, and little by little, I think we're winning." In the case of community-based education reform, dealing with the Naysayers will need to rely largely on demonstrating that they are either misinformed or are intentionally misrepresenting, are acting from a position of self-interest that is contrary to public well-being, and are simply wrong.

To do so requires that urgency be sustained through constant action and forward momentum. If neither occurs, the uncommitted will decide that this must not be very important after all. In the final section of her book, Wheatley asks, "Can I be fearless?" She observes that being fearless is not the same as being without fear, nor is it exactly the same as courage. It is not the spontaneous reaction to threat or danger, but the wise, studied encounter with our fears, and the decision that out of love—love for self, for our children, for our nation or posterity—we cannot allow fear to keep us from action. As we move forward fearlessly, fear melts away.

A Time for Fearlessness

I knew my maternal grandfather mainly as a big man with a crocked smile who took his sons, sons-in-law, and grandsons to Yellowstone each summer on a fishing trip and who, as a younger man, once swam across the Snake River at a particularly precarious spot on a dare. I knew in a vague way that he was Superintendent of Public Education for the State of Utah, an awareness that came primarily from family stories recounting the irony of his having run for the office against my father's father!

E. Allen Bateman was elected to the state's highest education office in late 1944 and, as he took office early the next year, realized that Utah's schools and colleges were not ready for the flood of young men who would soon be returning from World War II. During his first term he proposed a year-round school calendar for high schools, better utilization of facilities, a more flexible curriculum to accommodate high school students who preferred a vocation to a college-prep path, and strengthening of the state's junior college system. (News clippings indicate that he was concerned about the state's 75 percent high school graduation rate, and would undoubtedly be even more concerned to know that they were reported at 72 percent in 2006!)[11] Dr. Bateman also supported a constitutional amendment in the state to change the state superintendency from an elected position to one appointed by the State Board of Education. During the same 1948 election in which he won a second term, the constitutional amendment passed, and he was immediately appointed to the position by the State Board.

Also elected to state office in 1948 was Governor J. Bracken Lee, a self-made man who had not completed high school, and a hardcore fiscal conservative. Although the state had no bonded debt and a sizable surplus, Lee committed his administration to further reducing state spending, particularly in education—for which he seemed to have limited regard. He and my grandfather immediately clashed when Dr. Bateman recommended and received legislative approval for a $20,000 allocation to develop an office of research within the Department of Education to track school performance and facility utilization. Lee blocked the appropriation, but finally relented after Utah's Third District Court ruled that he had no authority as governor to veto the bill. But the battle lines were drawn, and an ongoing feud continued between the education establishment and the Governor until Lee was defeated in 1956.

In the history of J. Bracken Lee's political career in Utah, two episodes are always highlighted as symbolic of his controversial and colorful administrative style—the firing, while he was serving as Mayor of Salt Lake City, of the city's popular police Chief, W. Cleon Skousen, for raiding a gambling event at which Lee was present, and the withholding for a year the salary of E. Allen Bateman, the Superintendent of Schools. In family history, the latter is by far the more memorable. The old elected system for state superintendent had a statutory salary cap that was eliminated with the constitutional change, and the state board believed the old salary cap at $6,000 a year to be both inadequate and noncompetitive for the state's senior educator. As they prepared the 1951 budget proposal, the board voted to increase the superintendent's salary to $10,000 annually, and the legislature approved the appropriation. Rather than veto the bill, Lee worked through the state's Finance Commission to withhold payment, claiming the old salary cap was still applicable under the law. When in November of 1951 the State Attorney General ruled that the salary adjustment was legal, the Governor still refused to release any funds until an appeal again made its way through the state's courts. An article in the *Deseret News* in July of the following year reported that the superintendent had still not been paid and "had been forced to refinance his home and 'borrow money from friends' in order to eat and have a place to live."[12] The article reported that the State Supreme Court still had the case "under advisement."

After a year without a check, my grandfather did get his salary, paid off the second mortgage on his home, and settled his debts with friends. He died of a heart attack in 1960 while still in office, walking to a meeting of the Utah Education Association on Salt Lake's historic Temple Square. But in the intervening years, as we camped along the Snake River on our way to Yellowstone, not far from the site of "Grandpa's Great Swim," the subject of the year without a salary occasionally came up. One of my younger, less awestruck cousins once asked him, "Why didn't you just quit?"

I don't remember exactly what grandfather said, but the gist of it has stayed with me, and I suspect with all of my male cousins, ever since. As closely as I remember, he said "You don't just quit because someone is upset with what you are trying to do. If it's the right thing, you do it anyway. Sometimes it works out, and sometimes it doesn't. But you do it anyway."

For me, he was one of Margaret Wheatley's fearless men. He wasn't without fear, and he had matured well past the point of jumping into a swift current just because someone dared him to do it. He saw his state as being at a moment of crisis in education, and believed that in a thoughtful, systematic way he could do something about it—even when very powerful Naysayers were doing all in their power to oppose him.

In each community in America there are people like E. Allen Bateman—people of influence who believe that education is in a state of crisis, and that something can be done about it. They are people who know there will be opposition to any attempt at serious reform, but believe "you don't just quit because someone is upset with what you are trying to do." It is to these people that this book is written. In a very real sense, the future of our nation rests on your shoulders, but it will require every ounce of your resolve.

"The Hard is What Makes It Great"

In the 1992 film *A League of Their Own* about the women's baseball league created during World War II to fill the void left by the military draft of many major league ballplayers, there is a classic exchange between star player Dottie Hinson and her coach Jimmy Dugan. Dottie's husband has returned from the war, and she is considering leaving the league to return with him to Oregon. In his blunt, hard-edged way, Dugan reminds her that she loves baseball. That it has gotten inside her, and that it is what lights her up. Dottie's response is that at this point in her life she doesn't need his badgering, saying, "It just got too hard!"

"It's supposed to be hard," Dugan snaps back. "If it wasn't hard, everyone would do it. The hard is what makes it great!"

I would be less than honest if I suggested that leading the charge to reform a community school district isn't going to be hard. In fact, it will even be hard to work as one of the supporters, making calls to get people out to vote in a board election, attending meetings to speak up in support of a bold principal or superintendent, or even being firm with your child when she complains about homework or argues that he doesn't need to start second grade in a language immersion program or take another math class during his senior year. If it weren't hard, everyone would be doing it, and we wouldn't be plagued with declining performance and an expanding gap between the achievement levels of our students and the demands of an increasingly technical job market. But the reason we can still have great hope in our educational future is that there is nothing about the *hard* that virtually every one of us can't overcome. It doesn't require unusual physical strength or talent, and we don't have to be in a position of great authority to get it done. We must simply be fearless in the pursuit of excellence for our school district, and unwavering in our resolve.

By every measure, we are losing ground in the global education race, and government at the state and federal level is offering nothing that suggests substantive change. It is time for each of us to step forward and begin the process of rescuing America's Schools, beginning with our own community.

CHAPTER EIGHTEEN

The Rescuer's 12-Step-Guide to School Reform

Twelve Steps to School Reform

(1) Organize a community Committee on Reforming Education (CORE) Team.
(2) Initiate a community information campaign and establish a sense of urgency.
(3) Identify and encourage informed and committed citizens to run for the School Board.
(4) Initiate Legislation to extend charter privileges to all districts.
(5) Select school leaders who will courageously pursue improvement options available without charter legislation.
(6) Support the district in the selection, hiring, and evaluation of talented leadership.
(7) Insist on performance-based teacher evaluation, accompanied by increased compensation.
(8) Make mandatory a bachelor's degree in a content area for all teachers.
(9) Implement a rigorous faculty development program.
(10) Revise curriculum to insure a content-rich, challenging, course of instruction.
(11) Internationalize the curriculum and student opportunities.
(12) Provide a choice option for those who elect not to comply with new policies and standards.

This guide outlines a step-by-step path for leading a community through the process of bringing significant reform to its schools. It is certainly not the only path, and some communities may find that they are already well along it and can begin at some point farther into this Guide. I would caution, though, that if a proper foundation is not established through a program of community education and through the systematic placement of committed leadership on the school board and in the administration, actions taken might be more akin to "refreshing" an operating system on a computer that doesn't provide all of the programs and capacity the user requires, rather than installing a new system that accommodates every need. I have provided a rating scale beside each step in the process to encourage serious thought and discussion about the step, so that no critical element is overlooked or given too little attention. Some steps will take much longer to achieve than others, and some—such as modifying the interpretation of disability law—may never be accomplished. But, ideally, all steps should either be completed or underway if we are to achieve the greatest results.

(1) Organize a community Committee on Reforming Education (CORE) Team.

Those who have a compelling feeling that K–12 education needs to be reformed and find some resonance with what this book recommends should identify like-minded members of their community who will join them to create a CORE Team. Particular attention should be given to attracting people who have both influence and respect within the community, since part of the Team's responsibility is to identify and bring committed leadership to the School Board, and provide a program of community education that will raise the sense of urgency about the crisis.

A Core Team of at least twelve to sixteen members will facilitate the Team's work, with subgroups assigned to coordinate efforts in community education, school board development and liaison, legislative action, and financial support.

Low   High

Success in Accomplishing this Objective 1 2 3 4 5

(2) Initiate a Community Information Campaign and establish a sense of urgency.

The Community Education subcommittee will serve as the public information group for the CORE Team, collecting and disseminating information about district performance, national and international trends, challenges in education, security and the economy, and model programs that are addressing these challenges. This group should organize visits for CORE Team members to successful charter and non-charter public schools, invite outside specialists to the community to address the crisis in education and potential solutions, and arrange to get key community leaders who are not on the Team to places, conferences, and presentations that will highlight the need for education reform. This committee's role is to establish and maintain a sense of urgency.

Low   High

Success in Accomplishing this Objective 1 2 3 4 5

(3) Identify and Encourage Informed and Committed Citizens to run for the School Board.

The School Board Development and Liaison subcommittee's responsibility is to identify citizens who are well-informed about the national and international challenges resulting from our weakened position in education, and who will be fearless in pursuing appropriate remedies. These candidates should be vetted by the committee to the point that no obvious background issues will provide the Naysayers with fodder for a character assassination campaign, and ideally they should be free of personal connections with school district personnel that might create conflicts of interest as key decisions arise. In concert with the Community Education committee, this group works on planning and coordinating school board campaigns to seat selected members, then supports these members as changes in school policy are pursued.

                                                    Low   High
Success in Accomplishing this Objective   1   2   3   4   5

(4) Initiate efforts through the State Legislature to extend privileges provided by model charter school legislation to all districts in the state.

The CORE Team's Legislative Action subcommittee should research the states charter school provisions and those provided by other states that might provide better models. Legislative champions in both parties should be solicited, asking for their assistance in both drafting and supporting legislation that will extend charter school privileges to all school districts in the state, along with specific language requiring demonstrated improvement in key performance indicators if these freedoms are to be maintained. The Legislative Action Committee should attempt to find teams in other districts who are working on similar legislative change to launch a coordinated effort to get as many cosponsors for the bill as possible. (A possible strategy might be to send potential sponsors a synopsis of key chapters in this book, indicating that it will be followed by a call for an appointment to discuss the points in further detail and ways in which the legislator can assist with advancing the cause of education reform in the state.)

                                                    Low   High
Success in Accomplishing this Objective   1   2   3   4   5

(5) Select school leaders who will courageously pursue improvement options available without charter legislation.

The CORE Team and its supporters cannot wait for charter legislation or a restructured school board to begin encouraging school leadership to take corrective action. The Team should prioritize a list of initiatives, beginning with those that can most easily be achieved, and take its case to the board and administration. Some care could be taken to avoid recommendations, such as an extended school year, that might generate broad community concern, until a supportive school board and committed administrators can be placed, but policies that should generate broad public support, such as performance-based evaluation for teachers, strict disciplinary policies, and increased hiring standards for faculty, should be pursued immediately. If the board and administration are resistant, their hesitance arms the CORE Team with information that can be used to support reform board candidates, and insist on administrative changes.

                                        Low   High
Success in Accomplishing this Objective   1   2   3   4   5

(6) Support the School Board in the selection, hiring, and evaluation of talented, visionary, and courageous leadership.

Selection of school leadership is the responsibility of the board, but the CORE Team and its supporters can play a key role in assisting with development of job descriptions, criteria for evaluation, and support for difficult decisions once new leadership is in place. In concert, the board and administration shape policy, and it is the responsibility of the administration to exercise that policy within the parameters and to the level of expectation set by the board. The temptation to promote a faithful assistant superintendent or to settle for a weak applicant pool is seductive, and the CORE Team can serve as conscience to the board to insist on quality in the hiring process, support talented administrators, and protect them in times of controversy.

                                        Low   High
Success in Accomplishing this Objective   1   2   3   4   5

(7) As Board and Administrative Support are established, insist on performance-based teacher evaluation and better compensation for talented teachers.

This most critical step in supporting improvement in student achievement is to improve teaching, and that must begin by fairly and objectively evaluating teachers based on student-learning outcomes and rewarding outstanding performance accordingly. The CORE Team and its community support base must work with school and board leaders to implement a student evaluation program at every grade level that provides standardized measures of achievement to look at each student's progress from grade to grade. The *Teach for America* evaluation standards reviewed in chapter 10 provide a good model, and fairly evaluate teachers based on student improvement while those students are with the teacher, rather than solely on an inflexible grade standard. One of the CORE Team's, and the Board's, greatest responsibilities is to convince the public that we will get the quality of instruction we are willing to pay for, and if we want nationally and internationally competitive schools, teaching must become an attractive professional option to our most capable college graduates. If there is one public service that is worth an increased investment at the local level, it should be educating our children.

Low   High
Success in Accomplishing this Objective   1   2   3   4   5

(8) Make mandatory a bachelor's degree in a subject area for all teachers hired into the district.

The success of the European and Asian education systems is due in large part to high-achieving teachers who are schooled in the disciplines they teach. As part of the public appeal for greater teacher pay, the CORE Team and its supporters should work with school leaders to change policy to require all entering-teachers to have a baccalaureate in a discipline, and require all current teachers who are not within five years of retirement to obtain a degree in a content area. Teachers who choose not to pursue the credential should be willing to sign an agreement that they will not pursue a contract with the district beyond the fifth year. This increase in expectations should be accompanied by the increases in teacher compensation mentioned in Step 7, and one of the principal values and objectives of the CORE Team must be to increase teacher pay to enable the district to attract the best candidates and to make teaching a sought-after profession. A community that is not willing to pay more for excellent instruction should not expect to get it.

Low   High
Success in Accomplishing this Objective   1   2   3   4   5

(9) Implement a rigorous faculty development program that provides intensive summer training for new teachers, and continuing education for experienced teachers in their disciplines and in global issues affecting their students' futures.

A robust summer school program achieves several important improvement objectives, in that it affords students opportunities to remediate deficiencies, enjoy learning experiences in content areas in a less traditional format, and provides teachers with opportunities to experiment with learning strategies and ideas that the standard schedule doesn't afford. As *Teach for America* has demonstrated, it can also be the perfect "training camp" for novice teachers, working under the mentorship of an experienced and successful teacher. But continuing education, including opportunities to travel and study abroad, should be afforded to all teachers based on proposals that are tied specifically to the teacher's content area or to experiences that will contribute to legitimate professional growth. Education often overlooks the fact that its most significant capital resource investment is in its faculty, and continuous development of this resource must be a major priority.

The role of the CORE Team Financial Support Committee is to assist a local district foundation with generation of an endowment, or to create a foundation if one does not exist. Specific funds should be created within the foundation that support development areas such as international study opportunities for faculty, avoiding public criticism from the Naysayers that district public dollars are being spent on "teacher vacations."

	Low			High	
Success in Accomplishing this Objective	1	2	3	4	5

(10) Revise curriculum to insure that, in all major subject areas, students receive a content-rich, progressively challenging, and globally competitive course of instruction.

It may seem unusual to place teacher training, evaluation, and professional development ahead of curriculum revision. But as chapters 10 and 11 have illustrated, a great teacher is much more critical to student achievement than is a great curriculum. Good teachers find ways to compensate for weak content. But as teacher selection and preparation are being improved, the CORE Team should work with school leadership to appoint a community advisory committee on curriculum, consisting of business and civic leaders, faculty and administrators from area colleges, and ex-officio content specialists from beyond the region who might agree to review and comment on recommendations. This committee should be advisory, but charged with helping faculty design a curriculum by content area that will prepare students to compete at a level equal to or greater than their national or global competitors, with language instruction beginning at the elementary grades and four full years of instruction in mathematics, the sciences, humanities, social sciences, and the arts. For those students choosing to pursue a career in the vocational and technical areas, an equally rigorous applied option should be available, but with similar academic expectations through grade ten.

Within the context of curriculum revision, the board, administration and advisory groups (including faculty) should consider what daily schedule and calendar arrangement best facilitates learning outcomes, utilizing the best research available on the relationship between length of day and year, and student performance.

	Low				High
Success in Accomplishing this Objective	1	2	3	4	5

(11) Internationalize the curriculum and student opportunities.

Critical and analytical thinking require the ability to place ideas within a variety of contexts, and the broader and more varied the understood contexts, the greater the ability to imagine and create new possibilities. With work moving to Asia and Latin America and population centers shifting to Brazil, India, Nigeria, and Bangladesh, students must be familiar with other nations and their roles in the developing world if we are to relate to them effectively. As curricula are revised and as opportunities beyond the classroom are developed for students, a principal concern must be the development of a much deeper understanding of the world and our position in it. In the ideal world, every student during the summer, between the junior and senior year, would have an opportunity to spend three to five weeks in a language immersion program in a country appropriate to earlier language study, accompanied by a combination of experienced faculty mentors and novice teachers who are themselves expanding their global horizons.

	Low				High
Success in Accomplishing this Objective	1	2	3	4	5

(12) Provide a choice option within the district for those who elect not to comply with the rigors of the standard curriculum and school policy.

The model in the typical district with one or more charter schools is to provide a traditional school program, with the charter serving as the "choice" alternative. The model outlined in this book "charters" the district, allowing it to implement programs and policies that raise standards and performance expectations, increase teacher qualifications, alter the calendar and school day, and impose strict behavioral policies. Under this model, the chartered school becomes the new standard, and some provision is made for those who choose not to reach for these standards. Chapter 13 stressed that directing students along different academic paths must be based on interest and effort, not ability, and that two "tracks" should exist within the new chartered curriculum: one for those seeking college readiness leading to a baccalaureate degree or beyond, and another for those seeking additional training in a career or technical field for which a bachelor's is not required. The district must also develop or expand a non-charter option for those who elect not to test themselves against the rigors of the new expectations. As long as a student attends, works hard, and applies whatever talents he or she has to the chartered tracks, the student should be allowed to continue, understanding that the degree of success will depend on the degree of ability and application. But in order to insist on excellence in the reformed program, the district must provide a third option.

As this option is developed, the district and its leadership must establish consistent and defensible standards for determining when a student is placed in the alternative option, and when the presence of a disabled student in a mainstreamed or inclusive setting creates a restrictive environment for other students in the class. Clearly established policy, fairly and religiously followed, will be critical to making an "internal choice" model work.

	Low				High
Success in Accomplishing this Objective	1	2	3	4	5

Ideally, this 12-Step-Guide should have only 11 steps and would not begin with creating a CORE Team. When a genuine and enduring culture of education surrounds each public school in the United States, none will be required. I have intentionally avoided being too critical of elected school boards—despite the feeling of some respected reformists that as long as they govern our public schools, serious reform is unlikely or impossible. I see the chances that elected school boards will be replaced as being highly unlikely, and choose to believe instead that when the public becomes aware of the realities of our competitive position and its long term implications on our economy and national security, it will demand better results from the governance system we now have in place. Out of shame, obligation, or a renewed sense of responsibility, school boards will respond. In our founding years as a nation we were able to find a dozen brilliant and courageous men who created this system of citizen governance in cities of 5,000 inhabitants, like Boston and New York, and in the colonies largest city of Philadelphia that then boasted 25,000 residents.[1] Surely there are six to ten men and women in each city across America who can continue this proud tradition. But I am not firm enough in this conviction to leave either public outrage or duty-driven board response to chance, and the CORE Team serves the purpose of guiding communities toward both.

On the day I finished the first draft of this closing section, the *London Evening Standard* announced that Goldman Sacks had upgraded to 2027 its estimate of when China will become the largest economy in the world—fourteen years ahead of its earlier estimate and only one generation of school children from the time of this writing.[2] China should not be viewed as the Bogeyman, but as our global CORE Team—giving us every notice that our schools are not competitive, and without change we are failing as a nation. Their message to us is that there is no time to waste, and whether it be school boards, city councils, or local CORE Teams, some group of concerned citizens in every community in America must come forward to lead us into an education Renaissance—lest we don't get another opportunity.

NOTES

Introduction

1. Tom Friedman, "The New Untouchables," *The New York Times*, October 20, 2009.
2. Ibid.

1 We Need School Reform—and Soon!

1. National Center on Education and the Economy, *Tough Choices or Tough Times, A Report by the New Commission on the Skills of the American Workforce, 2007.*
2. David Hunn, "Worst Schools are in 6 City Areas," *St. Louis Post-Dispatch*, September 1, 2009, 1–2.
3. National Commission on Excellence in Education, "A Nation at Risk," April 1983, http://www.ed.gov/pubs/NatAtRisk/risk.html (accessed March 30, 2008), 1.
4. Ibid., 2.
5. Ibid.
6. Ibid.
7. U.S. Department of Education, "A Test of Leadership: Charting the Future of U.S. Higher Education," *A report by the commission appointed by Secretary of Education Margaret Spellings*, 2006.
8. Eric A. Hanushek and Ludger Woessmann, "The Role of Cognitive Skills in Economic Development," *Journal of Economic Literature* 46, 3 (2008): 607– 668.
9. Paul Klugman, "The Uneducated American," *The New York Times*, October 9, 2009, A-31.
10. Digest of Educational Statistics, 2007 Tables and Figures, "Averaged Freshman Graduation Rates for Public Secondary Schools, by State: Selected Years, 1990–91 through 2004–05." http://nces.ed.gov/programs/digest/d07/tables/dt07_102.asp (accessed April 1, 2008).
11. ETS was at one time the Educational Testing Service, best known as developers of the SAT exams. When it expanded the services it provided to areas beyond test development, the name was changed to simply ETS.
12. Educational Testing Service Policy Evaluation and Research Center, *America's Perfect Storm: Three Forces Changing Our Nation's Future*, January 2007, 3.
13. In the City of St. Louis where I live, three of the major high schools show graduation rates below 40 percent in one or more of the past five years, as shown on the State Department of Elementary and Secondary Education Web site.
14. Interactive Illinois Report Card, "Tomorrow's Builders Charter School," http://iirc.niu.edu/ (accessed April 6, 2009).
15. James B. Hunt and Thomas J. Tierney, "American Higher Education: How Does It Measure Up for the 21st Century?" The National Center for Public Policy and Higher Education, 2006, 8.
16. The College Board, "Average Mean Scores." http://www.collegeboard.com/student/testing/sat/scores/understanding/average.html (accessed March 28, 2009).
17. Edward Reingold and Peter Drucker, "Facing the 'Totally New and Dynamic,'" *Time Magazine*, January 22, 1990.

18. "A Test of Leadership," 1.

19. Hunt and Tierney, "American Higher Education," 8.

20. Sanford Bridge Report. "Betraying the College Dream" 2004, http://www.stanford.edu/group/bridgeproject/betrayingthecollegedream.pdf#search=%2212%20ways%20to%20be%20successful%20in%20college%22 (accessed June 30, 2009).

21. American Association of Community Colleges, "Fast Facts," 2006, http://www.aacc.nche.edu/Content/NavigationMenu/AboutCommunityColleges/Fast_Facts1/Fast_Facts.htm (accessed September 30, 2007).

22. Michael Lawrence Collins, "Setting up Success in Developmental Education: How State Policy Can Help Improve Community College Student Outcomes," June 2009, http://www.jff.org/Documents/AtD_brief_success_060809.pdf (accessed July 1, 2009).

23. Byron McClenney, "The Future of Developmental Education in Texas Community Colleges," (presentation to the Texas Commissioner for Higher Education, Austin, TX, September 29, 2008).

24. Achieving the Dream Data Notes, "February 2006: Developmental Math," http://www.achievingthedream.org/_images/_index03/February_2006_Developmental_Math.pdf (accessed July 3, 2009).

25. Nancy R. Lockwood, "The Aging Workforce: The Reality of the Impact of Older Workers and Eldercare in the Workplace," *HR Magazine*, December 2003.

26. Mitra Toossi, "Labor Force Projections to 2016: More Workers in Their Golden Years," *Monthly Labor Review*, Department of Labor Statistics (November 2007).

27. National Center on Education and the Economy, *Tough Choices or Tough Times* (San-Francisco: Jossey-Bass, 2008).

28. Ibid.

2   Where Are We Doing It Right?

1. National Commission on Excellence in Education, "A Nation at Risk," April 1983, http://www.ed.gov/pubs/NatAtRisk/risk.html (accessed March 30, 2008).

2. Association of American Colleges and Universities, "College Learning for the New Global Century," The National Leadership Council for Liberal Education & America's Promise, 2006, vii.

3. Ibid.

4. City-Data.com, "Helena, Arkansas," http://www.city-data.com/city/Helena-Arkansas.html (accessed December 31, 2008).

5. Economagic.com, "Unemployment Rate: West Helena, AR Micropolitan Statistical Area, Arkansas; Percent; NSA," http://www.economagic.com/em-cgi/data.exe/blsla/lauMC05483403 (accessed December 31, 2008).

6. KIPP: Delta College Preparatory School, "Percent Students Scoring Proficient or Advanced on Arkansas Benchmark Test," http://www.deltacollegeprep.org/results.htm (accessed June 18, 2009).

7. KIPP, http://www.kipp.org (accessed June 18, 2009).

8. Noble Street Charter School, "Academic Results," http://www.noblenetwork.org/AchievementsResults/AcademicResults/tabid/101/Default.aspx (accessed November 11, 2008).

9. Lou Ransom, "Several Chicago Named to Poor Performing List," *Chicago Defender*, March 25, 2009, www.chicagodefender.com/article-3596-several-chicago-schools-named-to-poor-performing-list.html (accessed July 1, 2009).

10. Jim Horn, "The Kult of KIPP: An Essay Review," *Education Review* 12, 9 (2009): 2.

11. Caroline M. Hoxby, S. Murarka, and J. Kang, *How New York City's Charter Schools Affect Achievement*, The New York City Charter Schools Evaluation Project. September, 2009, III-2.

12. Ibid., viii.

13. Ibid., V-5.

14. Lance T. Izumi and Xiaochin Yan, *Free to Learn: Lessons from Model Charter Schools* (San Francisco: Pacific Research Institute, 2001), 128.

15. Ibid., 129.

16. Chester E. Finn, Brunno V. Manno, and Gregg Vanourek, *Charter Schools in Action*. (Princeton, NJ: Princeton University Press), 2000, 71.

17. U.S. Charter Schools, "History." http://www.uscharterschools.org/pub/uscs_docs/o/history.htm (accessed December 31, 2008).

18. U.S. Charter Schools, "Charter Laws." http://www.uscharterschools.org/pub/uscs_docs/o/history.htm (accessed December 31, 2008).

19. KIPP Delta Public Schools, http://www.deltacollegeprep.org (accessed June 18, 2009).

20. School Board District Policies Manual Leopold R-3 School, "School District Philosophy," http://schoolweb.missouri.edu/leopold.k12.mo.us/policies.htm (accessed January 3, 2010).

21. Personal conversations with personnel and students in the Leopold R-3 School District, October 29, 2009.

22. Personal conversation with Mirra Anson, October 30, 2009.

23. Bill Gates, "2009 Annual Letter." http://www.gatesfoundation.org/annual-letter/Documents/2009-bill-gates-annual-letter.pdf (accessed August 10, 2009).

3 Where Reform Won't Happen

1. Barack Obama, Nationally Broadcast Press Conference, Washington DC, June 22, 2009.

2. Machiavelli, Niccolo, *Machiavelli's the Prince* (New York: Palgrave Macmillan, 1969).

3. "10 'O'Clock News Roundup – Hour 1." The Diane Rehm Show, NPR Radio, June 19, 2009. Transcribed from the Audio.

4. Taibbi, Matt, "Sick and Wrong," *Rolling Stone*, September 3, 2009, 58.

5. Kingdon, John W., *Agendas, Alternatives, and Public Policies* (Boston: Little Brown, 1984).

6. Rudolph, Frederick, *The American College and University* (Athens: University of Georgia Press, 1990).

7. Earle, Alice Morse, *Child Life in Colonial Days* (New York: Macmillan Company, 1935).

8. Constitution of the United States, Amendment X.

9. Gladwell, Malcolm, *Outliers* (New York: Little, Brown and Company, 2008).

10. Howard, Phillip K., *The Death of Common Sense* (New York: Warner Books, 1994), 11.

11. Diane Ravitch, "Privatization Will Not Help Us Achieve Our Goals: An Interview With Diane Ravitch." http://learningmatters.tv/blog/op-ed (accessed August 5, 2009).

12. Ibid.

13. Lance T. Izumi and Xiaochin Yan, *Free to Learn: Lessons from Model Charter Schools* (San Francisco: Pacific Research Institute, 2001), 20.

14. Coffman, James M., e-mail message, April 17, 2009.

4 Getting Someone to be Responsible

1. Ying Ying Yu, "A Duty to Family, Heritage and Country," *This I Believe*, National Public Radio, July 17, 2006. Transcript taken from http:www.npr.org/templates/story/story.php?storyId=5552257 (accessed July 25, 2006).

2. *Li Chi*, XXVII.

3. *The Analects of Confucius*, trans. Arthur Waley (London: George Allen & Unwin, 1938) XVII. 8.

4. Howard, Phillip K., *The Collapse of the Common Good: How America's Lawsuit Culture Undermines Our Freedom* (New York: Ballantine Books, 2001), 199.

5. A Nation at Risk, 1, http://www.ed.gov/pubs/NatAtRisk/risk.html (accessed March 30, 2008).

6. Theodore J. Kowalski, *The School Superintendent: Theory, Practice, and Cases*, 2nd ed. (Thousand Oaks, CA: Sage, Publications Inc., 2006), 134.

7. Jane Elizabeth, "School Boards' Worth in Doubt," Pittsburgh Post-Gazette, November 30, 2003.

8. Texas Education Agency, "2007–2008 School Report Card," Ysleta ES, http://ritter.tea.state.tx.us/cgi/sas/broker (accessed October 15, 2009).

9. Ysleta Independent School District, "History," http://www2.yisd.net/education/components/scrapbook/default.php?sectiondetailid=24&sc_id=1129231703 (accessed October 11, 2009).

10. Ysleta Independent School District, "Board of Trustee – Core Beliefs 2008–2009," http://www2.yisd.net/education/components/scrapbook/default.php?sectionid=4 (accessed November 23, 2009).
11. Gib Garrow, Personal conversation, September 7, 2009.

5 Finding and Supporting Great Leadership

1. Chester E. Finn, Brunno V. Manno, and Gregg Vanourek, *Charter Schools in Action* (Princeton, NJ: Princeton University Press, 2000), 109.
2. Betsy Taylor, "Urban Superintendents Hard to Keep," *USA Today Online*, September 30, 2008. http://www.usatoday.com/news/education/2008–09-28-superintendent_N.htm (accessed August 18, 2009).
3. "Urban School Superintendents: Characteristics, Tenure, and Salary Fifth Survey and Report," *Urban Indicator* 8, 1 (June 2006): 2.
4. Lawrence W. Lazetto, "Revolutionary and Evolutionary: The Effective Schools Movement." http://www.edutopia.org/files/existing/pdfs/edutopia.org-closing-achievement-gap-lezotte-article.pdf (accessed August 17, 2009).
5. Eric W. Robelen, "Nuances of Principalship Explored," *Education Week* 29, 14 (December 9, 2009): 1–11.
6. "Education Leadership Policy Standards: ISLLC 2008." The Council of Chief State School Officers. http://www.ccsso.org/content/pdfs/elps_isllc2008.pdf (accessed August 17, 2009).
7. Neibauer, Michael, "Michelle Rhee: A Teacher at Heart," *The Washington Examiner* (June 13, 2007). http://www.examiner.com/a-777760~Michelle_Rhee__A_teacher_at_heart.html (accessed June 11, 2009).
8. Ibid.; "The New Chancellor: Mayor Fenty Makes Clear that It's Not Business as Usual for District Schools," *The Washington Post,* June 12, 2007. http://www.washingtonpost.com/wp-dyn/content/article/2007/06/11/AR2007061102012.html (accessed July 18, 2009).
9. "District of Columbia Schools Continue to Make Steady Gains." District of Columbia Public Schools Web site. http://dcps.dc.gov/portal/site/DCPS/menuitem.3d9831ab117a6a932c69621 014f62010/? vgnextoid=639afa5c266d1210VgnVCM100000b912010aRCRD&vgnextfmt=de fault (accessed August 18, 2009).
10. Peter F. Drucker, introduction to *Mary Parker Follett: Prophet of Management*, ed. Pauline Graham (Boston: Harvard Business School Press, 1996), 2.
11. Mary Parker Follett, *Dynamic Administration: The Collected Papers of Mary Parker Follett*, ed. Elliot M. Fox and L. Erwick (London: Pitman, 1973), 31.
12. Ibid., 67.
13. Lee Bolman and Terrence Deal, *Reframing Organizations: Artistry, Choice, and Leadership* (San Francisco: Jossey-Bass, 2003).
14. Michelle Polley, Personal correspondence, October 17, 2009.
15. John Carver, *Boards That Make a Difference: A New Design for Leadership in Nonprofit and Public Organizations* (San Francisco: Jossey-Bass, 2006).

6 Developing a World Class Curriculum

1. Ross Miller, "Lesser and Greater Expectations: The Wasted Senior Year and College-Level Study in High School." Association of American Colleges and Universities, April, 2001, 4.
2. Laurence Steinberg, *Beyond the Classroom: Why School Reform Has Failed and What Parents Can Do* (New York: Simon & Schuster, 1996).
3. Ibid.
4. Ibid., 18–19.
5. "Highlights from TIMSS 2007: Mathematics and Science Achievement of U.S. Fourth and Eighth-Grade Students in an International Context." National Center for Educational Statistics (Washington DC: U.S. Department of Education), December 2008. iii.
6. "Highlights from PISA 2006: Performance of U.S. 15-Year-Old Students in Science and Mathematics Literacy in an International Context," National Center for Educational Statistics (Washington DC: U.S. Department of Education), December 2007, iii.

7. "Why We're Behind: What Top Nations Teach Their Children But We Don't" (Washington DC: Common Core, 2009), 13.
8. U.S. Department of Education, National Center for Education Statistics, Private School Universe Survey (PSS), 2001–2002.; http://nces.ed.gov/programs/digest/d03/tables/dt126.asp (accessed July 17, 2009).
9. "Why We're Behind: What Top Nations Teach Their Children But We Don't" (Washington DC: Common Core, 2009), iii.
10. Ibid., 3.
11. Ibid., 7.
12. Ibid., 18, 34.
13. Ibid., 14.
14. Victor Bandeira de Mello, C. Blankenship, and D. McLaughlin, "Mapping State Proficiency Standards Onto NAEP Scales: 2005–2007," National Center For Education Statistics, Department of Education, Washington DC, October 2009. vii.
15. Julia Smearman, "The European Edge in Education: What the United States Can Learn," Summary of the proceedings of a Roundtable discussion sponsored by STAGE, July 16, 2009, Washington DC. Taken from the Woodrow Wilson International Center for Scholars' *Wilson Center On Demand* Web site, http://www.wilsoncenter.org/ondemand/index. cfm?fuseaction=home.read&mediaid=BD549E22-E6CE-517F-1B86AE53A109C909 (accessed August 24, 2009).
16. Ibid.
17. William H. Schmidt, R. Houang, and S. Shakrani, "International Lessons about National Standards." Thomas B. Fordham Institute. Michigan State University, May 5, 2009. (The six lessons learned are quoted verbatim from the study, beginning on page 2, with the accompanying description paraphrased.)
18. Ibid., 5.
19. Ibid.
20. The National Governors Association, The Council of Chief State School Officers, and Achieve, "Benchmarking for Success: Ensuring U.S. Students Receive a World-Class Education," Washington DC: National Governors Association, 2008.
21. "Assessment Standards in Flux," Education Week, online, http://www.edweek.org/media/ 2009/11/02/nces-cl.jpg (accessed November 3, 2009).
22. The National Governors Association, 5.
23. Association of American Colleges and Universities, "College Learning for the New Global Century," The National Leadership Council for Liberal Education & America's Promise, 2006, vii.
24. Decker F. Walker, *Fundamentals of Curriculum: Passion and Professionalism*, 2nd ed. (Mahwah, NJ: Lawrence Erlbaum, 2002), 97.
25. E.D. Hirsch, jr. "The Case for Bringing Content into the Language Arts Block and for a Knowledge-Rich Curriculum Core for all Children," AFT Publications. http://www.aft.org/ pubs-reports/american_educator/issues/spring06/hirsch.htm (accessed June 23, 2009).
26. "General School Information." Parker Core Knowledge Charter School. http://www.ckcs.net/ pages/info/info.html (accessed August 19, 2009).
27. "Core Knowledge Sequence at a Glance," About Core Knowledge. http://www.coreknowledge.org/CK/about/sequence_glance.htm (accessed August 16, 2009).
28. Bill Gates, "2009 Annual Letter." http://www.gatesfoundation.org/annual-letter/ Documents/2009-bill-gates-annual-letter.pdf (accessed August 10, 2009).
29. Lance T. Izumi and Xiaochin Yan, *Free to Learn: Lessons from Model Charter Schools* (San Francisco: Pacific Research Institute, 2001), 6.
30. Decker F. Walker, *Fundamentals of Curriculum*, 109.

7 Math as the Language of Science and Technology

1. Kurt Badenhausen, "Most Lucrative College Majors," *Forbes*, June 18, 2008. http://www. forbes.com/2008/06/18/college-majors-lucrative-lead-cx_kb_0618majors.html (accessed September 15, 2009).

2. Karin Fischer, "As U.S. Retrenches, Asia Drives Growth Through Higher Education," *The Chronicle of Higher Education*, October 9, 2009, 21.

3. "The Nation's Report Card, Mathematics 2007," The National Center for Educational Statistics, U.S. Department of Education, September 2007.

4. "Highlights from the Trends in International Mathematics and Science Study," National Center for Educational Statistics, U.S. Department of Education, December 2007.

5. Julia Smearman, "The European Edge in Education: What the United States Can Learn," Summary of the proceedings of a Roundtable discussion sponsored by STAGE, July 16, 2009, Washington DC. Taken from the Woodrow Wilson International Center for Scholars' *Wilson Center On Demand* Web site, http://www.wilsoncenter.org/ondemand/index.cfm?fuseaction=home.read&mediaid=BD549E22-E6CE-517F-1B86AE53A109C909 (accessed August 24, 2009).

6. William H. Schmidt, "Papers and Presentations, Mathematics and Science Initiative," Ed.gov. http://www2.ed.gov/print/rschstat/research/progs/mathscience/schmidt.html (accessed February 7, 2010).

7. Ibid.

8. Graham A. Jones, C. Langrall, C. Thornton, and S. Nisbet, "Elementary Students' Access to Powerful Mathematical Ideas," in *Handbook of International Research in Mathematics Education*, ed. Lyn D. English (Mahwah, NJ: Lawrence Erlbaum Associates, 2002), 115–116.

9. Robert Kaplan and E. Kaplan, *Out of the Labyrinth* (New York: Oxford University Press, 2007), 1–3.

10. Barbara Lontz, Personal conversation, October 9, 2009.

11. Bradley S. Witzel and Paul J. Riccomini, "Optimizing Math Curriculum to Meet the Learning Needs of Students." *Preventing School Failure* 52, 1 (2007): 14.

12. Richard S. Kitchen, J. DePree, S. Celedon-Pattichis, and J. Brinkerhoff, *Mathematics Education at Highly Effective Schools That Serve the Poor* (Mahwah, NJ: Lawrence Erlbaum Associates, 2007), x.

13. Ibid., 89.

14. "Achievement Effects of Four Early Elementary School Math Curricula," Institute of Education Sciences, February, 2009.

15. Kitchen, 83.

16. Agnes Blum, "Better Math Education Requires Higher Expectations, Too," *Boston Globe*, December 2002.

17. Ibid.

18. Kitchen, 78.

19. Richard Florida, "America's Looming Creative Crisis," *Harvard Business Review* 83, 4 (2005): 122.

20. Richard Levine, "The American University and the Global Agenda," Speech to the Foreign Policy Association, April 16, 2008. http://www.fpa.org/calendar_url2420/calendar_url_show.htm?doc_id=668198 (accessed September 21, 2009).

21. Laura Burns, P. Einaudi, and P. Green, " S&E Graduate Enrollments Accelerate in 2007; Enrollments of Foreign Students Reach New High," InfoBrief, The National Science Foundation, May, 2009, 2. http://www.nsf.gov/statistics/infbrief/nsf09314/nsf09314.pdf (accessed September 22, 2009).

22. Florida, 125.

23. Edward Reingold and Peter Drucker, "Facing the 'Totally New and Dynamic,'" *Time Magazine*, January 22, 1990.

24. J. Myron Atkins and Paul Black, *Inside Science Education Reform: A History of Curricular and Policy Change* (New York: Teachers College Press, 2003), 38.

25. Ibid., 169.

26. Ibid., 74.

8 Second Language from the Start

1. Mara Hvistendahl, "China Moves Up to Fifth as Importer of Students," *The Chronicle of Higher Education* 55 (2009): A1.

2. "China will be the World's Biggest Economy by 2027," *London Evening Standard*, November 17, 2009, 1.

3. Mark Milliron, "Ten Emerging Insights on Education, Innovation, Technology and Tomorrow" (presentation, annual convention of the Missouri Community College Association, Lake of the Ozarks, Missouri, November 4, 2009).

4. Ernest LaBelle, "The World is Your Main Street," *Forum* (Modern and Classical Language Association of Southern California) 12, 2 (January 1975): 4–5.

5. "Some Facts About the World's 6,800 Tongues," CNN.com/US, http://archives.cnn.com/2001/US/06/19/language.glance/index.html (accessed August 23, 2009).

6. H. Jerold Weatherford, "Personal Benefits of Foreign Language Study," ERIC Digest ED276305, October 1986.

7. Len Lazarick, "Three Community Colleges Making Global Connections," *Trustee Quarterly* 2 (1998): 9.

8. Francois Grin, "Language Planning and Economics," *Current Issues in Language Planning* 4, 1 (2003).

9. Bruce Brown, "Building Character Through Sports: Developing a Positive Coaching Legacy," http://www.respectsports.com/pdf/Youth%20Sports%20Charts.pdf (accessed August 24, 2009).

10. P.W. Armstrong and J.D. Rogers, "Basic Skills Revisited: The Effects of Foreign Language Instruction on Reading, Math, and Language Arts." *Learning Languages* 2 (1997): 20–31.

11. C.G. Carr, "The Effect of Middle School Foreign Language Study on Verbal Achievement as Measured by Three Subtests of the Comprehensive Tests of Basic Skills" [Abstract], *Dissertation Abstracts International, 1994*, A 55(07), 1856.

12. "A Public Relations Handbook for Foreign Language Teachers," *Foreign Language Post*, Oklahoma State Department of Education, 1982, 21–31.

13. "2008 College-Bound Seniors: Total Group Profile Report," *The College Board*, 2008, 6.

14. Ingrid Pufahl, Nancy Rhodes, and Donna Christian, "What We Can Learn from Foreign Language Instruction in Other Countries," *Center for Applied Linguistics*, ERIC Digest EDO-FL-06, September 2001.

15. Ibid.

16. Holly Jacobs, personal conversation, October 15, 2009.

17. Jennifer Cuevas, personal conversation, October 15, 2009.

9   "Let Me Show You The World"

1. Jean Piaget, *The psychology of intelligence* (New York: Routledge, 1963).

2. John Dewey, *Experience and education* (London: Macmillan, 1938).

3. J. Rentilly, "A Thirst for (Unnecessary) Knowledge," *American Way*, September 2009, 71.

4. Sir Frances Bacon, *Essays, Civil and Moral: And The New Atlantis* (BiblioLife, 2009), 128.

5. Ruth Fulton Benedict, *Patterns of Culture* (Boston: Houghton Mifflin Co, 1934), ch. 1.

6. "Final Report: National Geographic- Roper 2006 Geographic Literacy Study," *The National Geographic Education Foundation*, May 2006, 8.

7. "What is Middfest?," Middfest International, http://www.middfestinternational.org/about.php (accessed January 1, 2010).

10   Hiring, Developing, and Evaluating for Excellence

1. Michelle Obama, "America's Future Lies in Its Teachers," *U.S. News & World Report* 146, 10 (November 2009): 37.

2. Robert Gordon, T. Kane, and D. Staiger, "Identifying Effective Teachers Using Performance on the Job," The Hamilton Project, Discussion Paper 2006–01, The Brookings Institute, April, 2006, 7.

3. Daniel Weisberg, S. Sexton, J. Mulhern, and D. Keeling, "The Widget Effect: Our National Failure to Acknowledge and Act on Differences in Teacher Effectiveness," The New Teacher Project, 2009.

4. Steven Brill, "The Rubber Room: The Battle over New York City's Worst Teachers," *The New Yorker*, August 31, 2009.

5. Ibid.

6. Nicholas D. Kristof, "Democrats and Schools," *New York Times*, October 14, 2009, A-35.

7. Jason Song, "Firing Tenured Teachers Can be a Costly and Tortuous Task," *The Los Angeles Times*, May 3, 2009. http://articles.latimes.com/2009/may/03/local/me-teachers3 (accessed June 24, 2010).

8. Ibid.

9. Bill Gates, "Prepared Remarks by Bill Gates, Co-chair and Trustee," November 11, 2008. http://www.gatesfoundation.org/FileReturn.aspx?column=Speech_Transcript_CMS&headerkey=Speech&page=%2Fspeeches-commentary%2FPages%2Fbill-gates-2008-education-forum-speech.aspx&name=speech.html&returntype=htmlfile (accessed January 2, 2010).

10. Gwang-Jo Kim, "Expansion of Education in Korea: From Access/Coverage to Quality Education," Presentation to the World Bank, November 12, 2009.

11. Andrew Hargreaves and D. Shirley, "The Fourth Way of Change," *Education Leadership* 66, 9 (October 2008): 58.

12. Lesli A. Maxwell, "Human Capital Key Worry for Reformers," *Education Week* 28, 14 (December 3, 2009): 1, 13.

13. "Our Corps Members," TeachforAmerica, http://www.teachforamerica.org/corps/index.htm (accessed November 20, 2009).

14. Zeyu Zu, J. Hannaway, and C. Taylor, "Making a Difference: The Effects of Teach for America in High School," National Center for Analysis of Longitudinal Data in Education Research, April, 2007 (Revised March 2009), 3.

15. "Regents Study Reports Effectiveness of New Teachers," Report to Board of Regents, Louisiana State University, August 27, 2009. http://www.regents.state.la.us/pdfs/PubAff/2009/Value%20Added_08–27-09.pdf (accessed October 22, 2009).

16. Linda Darling-Hammond, D. Holtzman, S. Gatlin, and V. Heilig, "Does Teacher Preparation Matter? Evidence about Teacher Certification, Teach for America, and Teacher Effectiveness," *Education Policy Analysis Archives* 13, 42 (October 12, 2005): 1–48.

17. Stephen Sawchuk, "Growth Model," *Education Week* 29, 3 (September 16, 2009): 27–29.

18. Lesli A. Maxwell, "Human Capital Key Worry for Reformers," *Education Week* 28, 14 (December 3, 2008): 13.

19. Steven Brill, "The Rubber Room: The Battle over New York City's Worst Teachers," *The New Yorker*, August 31, 2009.

20. Stephen Sawchuk, "Growth Data for Teachers Under Review," *Education Week* 28, 9 (October 22, 2008): 1, 14–15.

21. Daniel Weisberg, S. Sexton, J. Mulhern & D. Keeling, "The Widget Effect: Our National Failure to Acknowledge and Act on Differences in Teacher Effectiveness," The New Teacher Project, 2009, 2.

22. Ibid., 27–30.

23. Mary Parker Follett, *Mary Parker Follett: Prophet of Management*, ed. Pauline Graham (Boston: Harvard Business School Press, 1995), 16–17.

11   Keeping Teachers Current, Enthusiastic, and Energized

1. Marilyn Cochran-Smith and Kenneth M. Zeichner, eds. *Studying Teacher Education: The Report of the AERA Panel on Research and Teacher Education*, The American Education Research Association (Mahwah, NJ: Lawrence Erlbaum Associates, inc., 2005), 652.

2. "Benedum 5-year Collaborative Teacher Education," *College of Human Resources and Education, The University of West Virginia,* http://hre.wvu.edu/academics/undergraduate_programs/benedum_ collaborative_5_year_teacher_education (accessed October 27, 2009).

3. "Teacher Preparation Program," *Rutgers Faculty of Arts and Science,* http://teacherprep.camden.rutgers.edu/ (accessed October 27, 2009).

4. "Training," *Teach For America,* http://www.teachforamerica.org/corps/training.htm (accessed November 11, 2009).

5. Stephen Biermann, "Perceptions of Effective Instruction: A Community College Student View" (PhD diss., University of Missouri-St. Louis, 2010).
6. Joel Foreman, "Game-based Learning: How to Instruct and Delight in the 21st Century," *EDUCAUSE Review* 39, 5 (2004): 50–66.
7. Mark Milliron, "Ten Emerging Insights on Education, Innovation, Technology and Tomorrow" (presentation, annual convention of the Missouri Community College Association, Lake of the Ozarks, Missouri, November 4, 2009).
8. Joel Foreman.
9. John Kotter, *A Sense of Urgency* (Boston: Harvard Business Press, 2008), 81–82.
10. Beth Janssen, personal correspondence, October 27, 2009.
11. Brian Croone, personal correspondence, November 7, 2009.
12. "Welcome to the Toyota International Teacher Program Wiki," *Toyota International Teacher Program*, http://www.toyota4education.com/pmwiki.php?n=Main.WelcomeToTheToyotaInternationalTeacherProgramWiki?from=Main.HomePage (accessed November 9, 2009).
13. John Kotter, "Author John Kotter on Urgency," *Amazon Video*, http://www.amazon.com/Sense-Urgency-John-P-Kotter/dp/1422179710 (accessed November 2, 2009).
14. Debra Viadero, "Turnover in Principalship Focus of Research," *Education Week* 29 (2009): 1–14.
15. Charles Taylor, ed. *Sayings of the Jewish Fathers.* II: 19 (Cambridge: Cambridge University Press, 1897), 40–41.

12   Reinterpreting "Least Restrictive Environment"

1. Section 504 of the *Rehabilitation Act of 1973*, as amended, 29 U.S.C. 794.
2. Gus Douvanis and D. Hulsey, "The Least Restrictive Environment Mandate: How Has It Been Defined by the Courts?" *ERIC Digest*, ED469442, August 2002.
3. Ibid.
4. GREER V. ROME, 950 F.2D 688 (1LTH CIR. 1991).
5. OBERTI V. CLEMENTON, 995 F.2D 1204 (3RD CIR. 1993).
6. SACRAMENTO V. RACHEL H., 14 F.3D 1398 (9TH CIR. 1994).
7. POOLAW v. BISHOP, 67 F.3RD 830 (9TH CIR. 1995).
8. LIGHT V. PARKWAY, 41 F.3RD 1223 (8TH CIR. 1994); CLYDE K. V. PUYALLUP, 35 F.3D 1396 (9TH CIR. 1997); HARTMANN V. LOUDOUN, 118 F. 3D 996 (1997); DOE V. ARLINGTON COUNTY, 41 F.SUPP 599 (ED. VA. 1999).
9. Personal correspondence with the special education teacher, September 17, 2009.
10. "High School Completion by Youth with Disabilities," *Facts from NLTS2*, November 2005, 3.
11. Personal conversation with "Melinda," October 14, 2009.
12. Personal conversation with "Janice," October 14, 2009.
13. Robert Tomsho, "Parents of Disabled Students Push for Separate Classes," *The Wall Street Journal*, November 27, 2007, A1.
14. Ibid.

13   Providing Choice in the Public Arena

1. "Requirements for Continuing Status as an A+ School," Report to the State prepared by North St. Francis County R-1 School District, March 30, 2002.
2. Ken Owen, Personal correspondence, November 4, 2009.
3. Chester E. Finn, Brunno V. Manno, and Gregg Vanourek, *Charter Schools in Action* (Princeton, NJ: Princeton University Press, 2000), 31.
4. "About KIPP: Five Pillars," KIPP, http://www.kipp.org/01/fivepillars.cfm (accessed November 4, 2009).
5. Chester E. Finn, Brunno V. Manno, and Gregg Vanourek, *Charter Schools in Action*, 97.

14   Legislating for Change

1. Link to history of education reform timeline, http://www.google.com/#q=history+of+
   state+reform+education&hl=en&tbs=tl:1&tbo=u&ei=FV3LSvLiFo7GMNj5_MMD&
   sa=X&oi=timeline_result&ct=title&resnum=11&fp=2cca7b2e99206b9c (accessed November
   11, 2009).
2. Alene Russell and M Wineburg, "Toward a National Framework for Evidence of Effectiveness
   of Teacher Education Programs," *Perspectives*, The American Association of Colleges and
   Universities (Fall 2007).

15   Managing with Data

1. "About Achieving the Dream," *Achieving the Dream*, http://www.achievingthedream.org/
   ABOUTATD/GOALS/default.tp (accessed November 27, 2009).
2. "Toyota Production System," *Toyota*, http://www2.toyota.co.jp/en/vision/production_sys-
   tem/jidoka.html (accessed September 12, 2009).
3. "Toyota Quality," *Toyota Motors Manufacturing Kentucky, Inc.* http://www.toyotageorgetown.
   com/qualdex.asp (accessed September 12, 2009).
4. "Community College Survey of Student Engagement," *University of Texas at Austin*, http://
   www.ccsse.org/ (accessed November 18, 2009).
5. "Welcome to the NCCBP Website," *National Community College Benchmark Project*, http://
   www.nccbp.org/ (accessed November 19, 2009).
6. Julie A. Marsh, J. Pane, and L. Hamilton, "Making Sense of Data-Driven Decision Making
   in Education: Evidence from Recent RAND Research," *Occasional Paper, RAND Education*,
   http://www.rand.org/pubs/occasional_papers/2006/RAND_OP170.pdf (accessed November
   20, 2009).

16   Building Successful Partnerships

1. Barak Obama, "Remarks by the President on the American Graduation Initiative" (Speech at
   Macomb Community College, Warren, Michigan, July 14, 2009).
2. "Toward Equalizing Opportunity," The President's Commission Higher Education for
   Democracy, Vol. I. Establishing the Goals, http://www.ed.uiuc.edu/courses/eol474/sp98/tru-
   man.html (accessed November 10, 2009).
3. "The Role of State Policies in Shaping Dual Enrollment Programs," Office of Adult and
   Vocational Education, U.S. Department of Education, http://www.ed.gov/about/offices/list/
   ovae/pi/cclo/dual.html (accessed October 3, 2009).
4. Richard DuFour, "What is a 'Professional Learning Community'?," *Educational Leadership*,
   May 2008, 6–11.
5. "High School to Community College: New Efforts to Build Shared Expectation," *EdSource
   Report* (Mountain View, CA: EdSource, Inc, November 2009), 11.
6. Pew Charitable Trust, *The Bridge Project: Strengthening K-16 Transition Policies* (Stanford, CA:
   Stanford Institute on Education Research, 1999).
7. "MSU's Arabic Language Instruction Flagship Program Partners with Dearborn Public
   Schools," *MSU News* (Michigan State University), September 25, 2009. http://news.msu.edu/
   story/6883/ (accessed November 28, 2009).
8. Charlene Nunley, M. Shartle-Galotto, and M. Smith, "Working With Schools to Prepare
   Students for College: A Case Study," *How Community Colleges Can Create Productive Collaborations
   with Local High Schools*, ed. Jim Palmer (San Francisco: Jossey-Bass, 2000).
9. Ibid., 63–64.
10. "Academic Initiatives and the MC/MCPS Partnership at Montgomery College," http://
    www.montgomerycollege.edu/Departments/mcmcps/ (accessed November 13, 2009).
11. "Montgomery County Unified School District, 'Guiding Tenets,'" *Annual Report on Our Call
    to Action: 2008*, vi.

12. Ibid., iii.

13. Ibid.

14. Ysleta Unified School District, "Statistics," http://www2.yisd.net/education/components/scrapbook/default.php?sectiondetailid=26& (accessed November 27, 2009).

15. Martin Luther King, Jr., commonly referenced quote, but without place or occasion of attribution.

16. Jennifer Brinkerhoff, "Assessing and Improving Partnership Relationships and Outcomes: A Proposed Framework," *Evaluation and Program Planning* 25 (2002): 215–231.

17. Paula Glover, "Shared Boundaries: A Case Study of the Development of Collaborative Relationships Between Closely Situated Public Community Colleges and Public Four-Year Institutions" (PhD diss., University of Missouri-St. Louis, 2008).

17 Sustaining and Enhancing Reform

1. Michael Keller, "FreeCell – Frequently Asked Questions (FAQ)," http://www.solitairelaboratory.com/fcfaq.html (accessed November 23, 2009).

2. Margaret Wheatley, *Turning to One Another: Simple Conversations to Restore Hope to the Future* (San Francisco: Berrett-Koehler Publishers Inc., 2009), 124.

3. Shirley Ann Jackson, "The Quiet Crisis: America's Economic and National Security at Risk." http://www.rpi.edu/homepage/quietcrisis/index.html (accessed November 15, 2009).

4. Wheatley, 126.

5. John Kotter, *A Sense of Urgency* (Boston: Harvard Business Press, 2008).

6. John Rassias, "A Philosophy of Language Instruction," The *Rassias Foundation*, http://www.dartmouth.edu/~rassias/john_rassias/about.html (accessed November 15, 2009).

7. Kotter, *A Sense of Urgency.*

8. Kotter, 147.

9. Ibid.

10. Ibid.

11. "Understanding High School Graduation Rates in Utah," Alliance for Education Excellence. http://www.all4ed.org/files/Utah_wc.pdf (accessed November 16, 2009).

12. "Utah Official Still Lacks Pay," *Deseret News*, July 15, 1952, 1.

18 The Rescuer's 12-Step-Guide to School Reform

1. Henry William Elson, *History of the United States of America* (New York: Macmillan Publishing, 1904), 198–200.

2. "China will be the world's biggest economy by 2027," *London Evening Standard*, November 17, 2009, 1.

INDEX